Luminous Phenomena
A Story of Spontaneous Combustion, Phosphorescence and Other Cold Lights

Luminous Phenomena
A Story of Spontaneous Combustion, Phosphorescence and Other Cold Lights

By

Allan Pentecost
King's College London
Email: allan.pentecost@kcl.ac.uk

ROYAL SOCIETY
OF **CHEMISTRY**

Hardback ISBN: 978-1-83707-002-2
EPUB ISBN: 978-1-83707-003-9
PDF ISBN: 978-1-83707-004-6

A catalogue record for this book is available from the British Library

The Royal Society of Chemistry is a charity, registered in England and Wales, Number 207890, and a company incorporated in England by Royal Charter (Registered No. RC000524), registered office: Burlington House, Piccadilly, London W1J 0BA, UK, Telephone: +44 (0) 20 7437 8656.

For further information see our website at www.rsc.org

For general enquiries, please contact books@rsc.org

For EU product safety enquiries, please email books@rsc.org or contact Royal Society of Chemistry Worldwide (Germany) GmbH, Römischer Hof, Unter den Linden 10, 10117 Berlin.

Printed in the United Kingdom by CPI Group (UK) Ltd, Croydon, CR0 4YY, UK

Preface

The British countryside can be depressingly dark and un-interesting on a cloudy and moonless night. There are no fire-flies to watch nor noisy insects to hear. There are owls and foxes of course but there is little to stimulate the eye apart from the lights of some lonely farmhouse, distant road or high-flying aircraft. But all is not lost and persistence can reward the night-walker with some illuminating surprises. For example some of the autumn fungi are bioluminescent and their hidden my-celium can often be seen glowing in dark woods when rotting wood is disturbed. One of the most spectacular but rare British sights is the glow worm. This is the female of a small beetle, *Lampyris noctiluca*, whose lower abdomen has large light-pro-ducing organs that are remarkable in being backed by small crystals that help to reflect the light and concentrate it in a particular direction. It is used to attract males, and the insect can reduce or extinguish the light if disturbed.

On a scout camp at Brightling in Sussex in the summer of 1961 we were about to bed down in our WW1 army tent when a commotion was heard outside. Two boys were talking excitedly, and it was not long before they entered and told us the news. Hundreds of glow-worms had surrounded the tent. They were everywhere! We dressed and ran outside. There in the long grass and low bramble bushes they made a spectacular sight scattered here and there over a large area but mostly on the grass stems.

Luminous Phenomena: A Story of Spontaneous Combustion, Phosphorescence and Other Cold Lights
By Allan Pentecost
© Allan Pentecost 2025
Published by the Royal Society of Chemistry, www.rsc.org

One patrol member had a bright idea and went off to get a jam jar. We gently transferred about 50 larvae and returned to our tent then listened to ghost stories recounted by our patrol leader under the eerie glow of our home-made lamp. It was a memorable night.

Along with glow worms, several other luminous phenomena described in this book are now uncommon or rare – at least they are seldom reported. It is mostly because of this, and the fact that these events are often of short duration and unpredictable in time and space, that little is known about them. Much remains to be learned at a time when we seem to know so much about everything else. This book cannot do justice to all the luminous phenomena that have been observed and written about and some selectivity has been necessary. It was considered important to include some well-known and much studied phenomena such as the combustion of fuels, lightning, auroras and even light-emitting diodes but these are added mainly to provide context to a wide range of less familiar phenomena such as St Elmo's Fire and bioluminescence.

The Introduction contains a brief resumé of the scientific basis of combustion and light emission. It provides some definitions of these and other terms that will be frequently encountered in the following chapters and will attempt to define the many types of luminescence that have appeared over the years beginning with incandescence. These terms were originally proposed to assist in an understanding of the various luminous phenomena although some have unfortunately overlapped with others often causing confusion. The Introduction also seemed the most appropriate place to investigate the nature and limitations of human vision. The human eye, one of the most remarkable products of nature, still has its limitations as most of us are well aware. It is necessary to put some figures to both its precision and sensitivity to the visible radiations as it remains the most important means of gathering scientific data on the emission, reflection and scattering of light in nature.

Where appropriate an historic approach has been taken in the succeeding chapters in order to follow the development and understanding of phenomena, often beginning with the earliest recorded literature. This approach should assist those readers whose science may be a little 'rusty' by introducing the topics gradually but I have tried to keep both the theory and 'jargon' of physics and chemistry to a minimum. I have tried to keep the text flowing and have avoided using too many formulae and unnecessary diversions. On the few occasions when recourse to theory has been unavoidable the reader is directed to other more appropriate and accessible works if required.

The second and third chapters deal with spontaneous combustion, beginning with the chemistry of the process '*in vitro*', that is to say its observation and experimentation within the laboratory environment. The third deals with spontaneous combustion in nature including such topics as the burning of compost piles, hayricks and oil shales, of living vegetation and animals including man himself. This topic is beginning to attract much attention owing to the recent dramatic increase in forest fires and their devastating consequences. Many of these phenomena remain little understood but some recourse has been made to the findings outlined in the previous two chapters. The next two chapters lead us into more physical phenomena. Chapter 4 deals with St Elmo's Fire. This is an electrical manifestation that has received much interest among seafarers for countless generations but nowadays more likely to be seen on the wingtips of jet liners and helicopters. A form of electroluminescence, its investigation played no small part in the history of science leading ultimately to an understanding of the nature of electricity.

Light emission by radioactivity is dealt with in Chapter 5. Are the unexplained lights sometimes seen by cavers in the Yorkshire Dales caused by leaks of radon gas from the underlying granite? This topic is explored with reference to several other forms of radioluminescence.

The next two chapters are devoted mainly to natural forms of luminescence. Chapter 6 investigates light emission by stones and powders including some famous gemstones and the remarkable Bologna stone of Italy. Also included are luminous paints used by the ancients and tenebrescent minerals that darken on exposure to sunlight. Several related phenomena are included here such as those produced by triboluminescence and thermoluminescence. Examples are drawn from earthquake lights, the fabled Hessdalen lights of Norway and the snapping claws of the pistol shrimp. In Chapters 7 and 8 are concerned with particular aspects of bioluminescence. This is an area of biology that has received a good deal of recent attention and from it, many useful applications have been made. For this reason the chapter focusses upon a small number of topics with links to some of the previous chapters. One example is bioluminescence in the open sea. Not only is this a phenomenon operating on a vast scale – thousands of square kilometres of the ocean are sometimes involved, but there are still many intriguing questions concerning its origins, its development and the reason for its very existence. The following chapter looks at a few examples of bioluminescence in the terrestrial flora and some unusual plants capable of reflecting sunlight by their lens-like cells.

This book concludes with a chapter delving into the mysterious Will o' the wisp or *ignis fatuus* to give its formal designation. This has long been an interest of mine and it still defies a fully rational explanation, largely owing to its rarity and erratic occurrence. British descriptions of the phenomenon are detailed although recourse to its worldwide distribution is also made. The reader will find here a number of historic accounts several of which are based upon folklore plus some more recent and previously unpublished sightings. The long regarded explanation as a form of spontaneous combustion has been questioned by many scientists but the reader might be sufficiently enthused to follow up on some of these unforgotten or untrodden pathways. There is here a good subject for the amateur scientist who is willing to beat a trail through dark untrodden mires. Some of the author's own experiences are here shared.

I hope that readers will find within this small book plenty of material to stimulate their curiosity. As other writers have expressed: curiosity is a willing, proud and eager confession of ignorance, but it is at the heart of what motivates us to understand our world.

Allan Pentecost

Acknowledgements

This book has developed slowly over the course of several de-
cades as time allowed. Throughout this period I had the support
and assistance of my wife and son, and during the earlier years,
my parents. All contributed to this work in one way or another,
from providing obscure sources of information to the patience
and encouragement required to keep a book of this nature on
track. Among the many other people who have given assistance,
special mention should be to friends and colleagues who have
taken interest in the work and provided many valuable suggestions and ideas. A good number of people, some of whom I did
not know personally have generously provided information,
photographs and images for the work. Two anonymous reviewers
also provided helpful suggestions. Others who have helped include friends and neighbours who read some of the draft
chapters giving useful feedback and advice. Finally I am in-
debted to the production team and editors at the Royal Society of
Chemistry for their patience, encouragement and assistance,
often at a moment's notice.

A. P.

Luminous Phenomena: A Story of Spontaneous Combustion, Phosphorescence and Other
Cold Lights
By Allan Pentecost
© Allan Pentecost 2025
Published by the Royal Society of Chemistry, www.rsc.org

Glossary

activation energy: the energy required to start a chemical reaction.

adiabatic: in gases a process or change where no heat is added or taken away from the system.

aliphatic compound: organic compounds that lack the stabilising effect of aromatic rings.

alpha-rays: fast-moving helium nuclei produced by some radioactive elements.

anhydrous: compounds without associated water molecules.

anode: the positive electrode in some electrical circuits. Negative ions and electrons are attracted towards it.

aromatic compound: hydrocarbons with a ring structure incorporating a special type of bonding increasing the stability and changing the electronic properties of the molecule.

atom: the smallest part of an element that can exist chemically

auto-ignition temperature: the lowest temperature to which a material will burn in air.

autokinesis: the illusion that a dim stationary light viewed in the dark appears to move of its own accord.

beta-rays: fast-moving electrons produced by some radioactive elements.

bond energy: the amount of energy associated with a chemical bond. It can be obtained from the heat of atomisation of a molecule.

Luminous Phenomena: A Story of Spontaneous Combustion, Phosphorescence and Other Cold Lights
By Allan Pentecost
© Allan Pentecost 2025
Published by the Royal Society of Chemistry, www.rsc.org

bremsstrahlung: radiation produced during the rapid deceleration of electrons and other charged particles to satisfy the law of energy conservation.

caramelin: coloured substances produced by the heating of sugars. They are a complex mixture of cyclic organic compounds.

carboxylic acid: an organic acid containing at least one CO–OH group.

catalyst: a substance that increases the rate of a chemical reaction without undergoing any permanent change.

cathode: the negative electrode in some electrical circuits. Positive ions are attracted towards it.

combustion: a rapid and self-supporting exothermic reaction usually accompanied by flame. It is often restricted to the burning of a fuel by the oxygen of the air.

cone (eye): specialised cells in the retina of the eye responsible for the sensing of bright light.

coulomb: the amount of electric charge transported in one second by a current of one ampere.

covalent bond: chemical bonds formed by the sharing of valency electrons between atoms.

cytoplasm: that part of a living cell that is found within the outer membrane (plasmalemma). It usually excludes the cell nucleus.

deflagration: an extremely rapid chemical reaction just falling short of an explosion.

DTA: analysis by heating a substance in an inert atmosphere and plotting its weight change with temperature.

electrolysis: chemical reactions produced by passing an electric current through an electrically conducting liquid. At the anode, negative ions may lose their electrons and become neutral species (an oxidation). At the cathode, chemical reduction occurs.

electron: a negatively charged particle that occurs in all atoms and exists within shells around the nucleus.

electron acceptor: an element or compound that will accept an electron from another element or compound.

electron volt: a small unit of energy. It is equal to the work done on an electron moving through an electrical potential difference of one volt. Symbol eV

electrophile: a molecule or ion that is electron deficient and accepts electrons. Many are reducing agents.

element: a substance that cannot be decomposed into simpler substances. In any element, all the atoms have the same number of electrons and protons.

exothermic reaction: a chemical reaction that releases heat into the surroundings.

facultative anaerobe: an organism that can respire in the presence of oxygen and in its absence can carry out fermentation.

Farbe centre: a region where an anion in a crystal lattice is replaced by an unpaired electron. This can cause colouration in transparent crystals due to selective light absorption by the centre.

fermentation: a form of respiration undertaken by microbes in the absence of molecular oxygen.

fluorescence: a type of luminescence where the excited electrons of an atom or molecule rapidly return to a ground state with emission of a photon. It most often occurs when a more energetic photon is absorbed by the atom/molecule.

fluorspar: the mineral form of calcium fluoride, CaF_2. It often forms large and colourful cubic crystals.

forbidden transition: transitions between energy levels in atoms and molecules that are not allowed in the quantum mechanical models for the electron. In practise, such transitions can occur but are rare.

free radical: atoms with an unpaired valency electron. Often formed by pyrolysis where a bond is broken in a molecule without the formation of ions. Most free radicals are chemically highly reactive.

halogen: one of the Group 17 elements with one electron short of a stable noble gas configuration. All are strongly electronegative and include chlorine and fluorine.

heat of formation: the energy liberated or absorbed when one mole of a compound is formed from its elements.

hydrolysis: the chemical reaction of a compound where one of the reactants is water.

incandescence: visible radiation emitted from a body subjected to a high temperature.

induction: in electrostatics, the production of electric charge on a conductor of electricity under the influence of an electric field.

ion: an atom or group of atoms that has either lost or gained electrons resulting in their becoming electrically charged.

ionic bond: a type of chemical bond where the valency electrons of one atom or group are transferred to another atom or group. The resulting bond is caused by electrostatic attraction.

ionisation energy: the minimum energy needed to completely remove an electron from its atom or molecule. The energy needed to remove the least strongly bound electron is called the first ionisation potential.

isotope: atoms of an element that have the same number of electrons and protons but different numbers of neutrons. Hydrogen has three isotopes with no (hydrogen), one (deuterium) or two (tritium) neutrons. The last is radioactive.

kilojoule (kJ): a unit of energy often used in thermochemistry.

luminescence: emission of light from a substance without the use of heat.

mafic: rocks with a high content of ferromagnesian minerals. An example is basalt.

mole: the weight of a substance equal to its molecular weight. A mole of sodium chloride weighs 58.44 grams (atomic weights of sodium and chlorine are 22.99 and 35.45 respectively).

molecule: smallest part of a chemical compound that can take part in a chemical reaction.

molecular formula: a shorthand method of describing the composition of a molecule. The molecular formula of water is H_2O.

nanogram: a billionth of a gram $(10^{-9}$ g).

nucleophile: a molecule or ion that can accept electrons. Many are oxidising agents.

osmosis: the passage of water across a membrane from a solution containing a higher content of solids to a lower content of solids. The membrane is semi-permeable giving preference to the passage of water.

Pauli exclusion principle: no two electrons in an atom can share all four quantum states.

peroxy compounds: those with molecules containing the structure R–O–O–R. Both oxygen atoms are bonded to other groups (R) that are not necessarily the same. They are often strong oxidising agents.

phosphorescence: a type of luminescence where the excited electrons of an atom or molecule return to a ground state with emission of a photon. It most often occurs when a more energetic photon is absorbed by the atom/molecule. It is distinguished from fluorescence where the time delay is shorter owing to a different kind of electronic transition.

photon: a quantum of electromagnetic radiation. Its energy is proportional to the frequency of the radiation.

polycyclic: organic compounds with two or more ring structures formed with a carbon framework. They are often produced by plants.

proton: a positively charged particle found in the nucleus of an atom.

pyrolysis: chemical decomposition that occurs upon heating. Refers particularly to organic compounds.

pyrophore: a spontaneously combusting material. Most often refers to a chemical mixture that has been previously heated in the absence of air.

quantum: an indivisible quantity of energy.

redox: A term for oxidation-reduction reactions where an atom/molecule gains electrons (**red**uction) and another atom/molecule loses electrons (**oxi**dation).

redox potential: a measure of the tendency for a chemical species to gain or lose electrons to an electrode. It varies from about $+3$ to -3 volts. In ecology, environments where reducing conditions prevail have negative potentials.

reduction: a type of electron transfer process occurring during chemical reactions. A simple example is the reduction of silver chloride (AgCl) to silver metal. In this salt the silver exists as the positive silver ion (Ag^+). UV radiation can cause the salt to dissociate and the chloride (Cl^-) electron can return to the silver ion and form electrically neutral silver metal.

rod (eye): specialised cell in the retina of the eye responsible for the detection of low levels of light.

self-ignition: an alternative term for spontaneous combustion.

serpentinisation: the transformation of minerals olivine and pyroxene into serpentinite (serpentine) under the influence of water. During the process, some water is reduced to hydrogen.

sorption: the attachment or take-up of a gas or liquid to a solid. In some cases the gas/liquid becomes bonded to the solid, weakly or strongly dependent upon the materials and conditions.

spontaneous combustion: combustion of a material in the absence of forced ignition, *i.e.* a spark or flame.

symbiosis: association of dissimilar organisms to their mutual advantage.

terpene: a class of unsaturated hydrocarbons (*i.e.* with some double bonds) found in plants.

thermogenic: a material or compound generated with the aid of heat.

unit cell: the smallest part of a crystal structure that shows the entire framework of the crystal.

Contents

Luminous Phenomena: A Story of Spontaneous Combustion, Phosphorescence and Other Cold Lights
By Allan Pentecost
© Allan Pentecost 2025
Published by the Royal Society of Chemistry, www.rsc.org

CHAPTER 1

Introduction to Chemistry, Flames and Radiation

1.1 ATOMS, ELEMENTS AND COMPOUNDS

It was long suspected that the material world was composed of simple units of matter. When combined in different ways, these units could then be made to account for all the known materials. The Greeks believed these units to be earth, fire, air and water but they ultimately proved unsatisfactory and were based upon little by way of experiment and observation. Much closer to our own time, the real units were uncovered. Slowly at first, but then with increasing rapidity once chemists grasped the methodology of observation and experiment, the units were revealed in what may be called the first Golden Age of Chemistry. It began in the 1770s with the discovery of oxygen, and in the following 150 years, most of the units were discovered. Of these only 83 are believed to be 'stable' that is, to have existed unaltered throughout the history of the Earth, the remainder being un-stable and radioactive. These units are now known as the elements. Only about 20 are at all common in the Earth's crust, although some of the rarities such as gold and silver are well known to us as valued items. Most elements are metals and look much the same – it's not often possible to distinguish one

Luminous Phenomena: A Story of Spontaneous Combustion, Phosphorescence and Other Cold Lights
By Allan Pentecost
© Allan Pentecost 2025
Published by the Royal Society of Chemistry, www.rsc.org

metallic element from another. The same cannot be said about the non-metals, some of which vary greatly in their colour and form.

It's now time to look at some important definitions. An atom is the smallest part of an element that can exist chemically. Atoms combine in definite proportions with other atoms to form compounds. Molecules are the smallest parts of chemical compounds that can take part in a chemical reaction. For example, carbon dioxide is a compound. This molecule contains one atom of carbon and two atoms of oxygen.

Great advances in the understanding of the elements and their compounds began in 1897 with the discovery of the electron by J. J. Thompson. Theorists had previously held the view that atoms were indivisible and had a range of shapes. Hydrogen atoms were regarded as spherical while oxygen atoms were thought to be doughnut-shaped with a hole large enough to admit two hydrogen atoms. This, it was argued, would explain the structure of the water molecule, H_2O. However the model was soon found to be inadequate as more complex molecules were discovered. Thompson worked with cathode ray tubes. These are tubes of glass that have two electrodes inserted into them for the passage of an electric current. When a tube is evacuated and a voltage is applied particles were seen to travel from one electrode to the other. He was able to show that these particles were much smaller than atoms and also carried an electric current. He discovered that the atoms themselves are divisible into smaller units. Thompson called these particles *corpuscles* but was later persuaded to rename them *electrons*. This discovery opened the doors to the structure of atoms and their relationship to each other. A hydrogen atom for example contains one electron which is negatively charged and one proton which is positively charged. The two charges cancel and the hydrogen atom is said to be neutral. Electrical neutrality is a feature of all atoms. In cases where atoms have an unequal number of electrons and protons they are said to be charged and are called ions. Elements differ in the number of electrons they contain. As mentioned above, the element hydrogen has one electron and is described as having the atomic number 1. Each element has a different atomic number. The atomic number for carbon is 6 and that of oxygen is 8.

One of the most familiar elements is iron. A small piece of iron, such as a nail, is composed of myriads of iron atoms. If an iron nail, 30 mm long, is divided into two, and one half divided again and then again, the individual atoms would eventually be revealed. Atoms are so small that we would need to make 78 repeated divisions to achieve this. Remarkably, structures the size of atoms can actually be revealed with special forms of microscope. Atoms were believed to exist long before they were seen with these microscopes. Studies on the theory of gases led the Austrian scientist Johann Loschmidt to estimate the size of the nitrogen (N_2) and oxygen (O_2) molecules making up the air. This in turn allowed the number of molecules to be found in a cubic centimetre of air at room temperature. This number is called the Loschmidt constant and it's large, 2.65×10^{19}. In other words ten multiplied by itself 19 times and then multiplied by 2.65.

When we distinguish the different types of element, a shorthand method of writing is used. The element iron, and also its atom is written as Fe, confusing in this case because it would seem more logical to write it as Ir. Historically, iron, along with some other familiar elements has its shorthand derived from the Latin, in this case, ferrum. The symbol Ir is used for another element, iridium.

The atoms of most of the elements are not found isolated from one another in our immediate surroundings. Instead they prefer to be joined together as molecules. For instance, oxygen (chemical symbol O) usually consists of pairs of atoms and the resulting molecule is written as O_2. It is sometimes referred to as dioxygen. Virtually all of the elements can bond with other elements to give molecules. The number of potential combinations is huge. If the simplest of all elements, hydrogen, could bond with the remaining stable elements (elements that are not radioactive) there would be at least 82 different types of molecule. In some cases more than one atom of hydrogen can be bonded to another element. In the case of carbon hundreds of different molecules can be formed with hydrogen and they are called hydrocarbons. When hydrogen combines with the metallic elements the molecules are called hydrides and some of these will be encountered later. With helium (He), the next simplest element with atomic number 2, the situation is different. It will not combine with other elements at all, even with itself. So helium exists as single atoms. There are several other unreactive elements like helium and they are called the noble gases. They include one called radon (Rn) which we shall also encounter later.

The elements therefore appear to have differing bonding tendencies. Some like hydrogen and iron bond with other elements willingly, but with others like the noble gases, bonding is more difficult, or never observed. Even if we accept that not all combinations of elements are possible, it still leaves a large number that are.

A familiar molecule where three atoms belonging to two elements are bonded together is water, H_2O. Chemists soon found that more complex combinations could be formed with three or more elements. For example, sulphuric acid, used in car batteries, contains the elements hydrogen, oxygen and sulphur (S) and is written H_2SO_4. These further combinations lead to literally millions of known compounds, with many others still awaiting discovery. You may be relieved to find that the number of compounds appearing in this book is small and rarely involves more than three elements. It is an interesting fact that among the huge number of known compounds few contain more than five different elements. This includes most of the compounds making up all living things.

Once chemists had discovered fifty or so elements, patterns began to appear. It seemed that some of the elements were related to one another through their chemical and physical properties. Many attempts were made to clarify these relationships and finally in 1869, the Russian chemist Dimitri Mendeleev managed to arrange them in columns and rows according to their atomic number. The result was the Periodic Table. The table takes several forms and one is shown in Figure 1.1. The columns of this table are termed *groups* and they number 18 running from left to right. The rows, fewer in number, are called *periods* and are numbered from top to bottom. This allows each element to be uniquely described in terms of its group and its period. Moving from left to right and then down, in the manner of reading a book, the atoms of these elements become increasingly large along with their atomic number. At each step, the atomic number of the atom (*i.e.* the number of electrons)

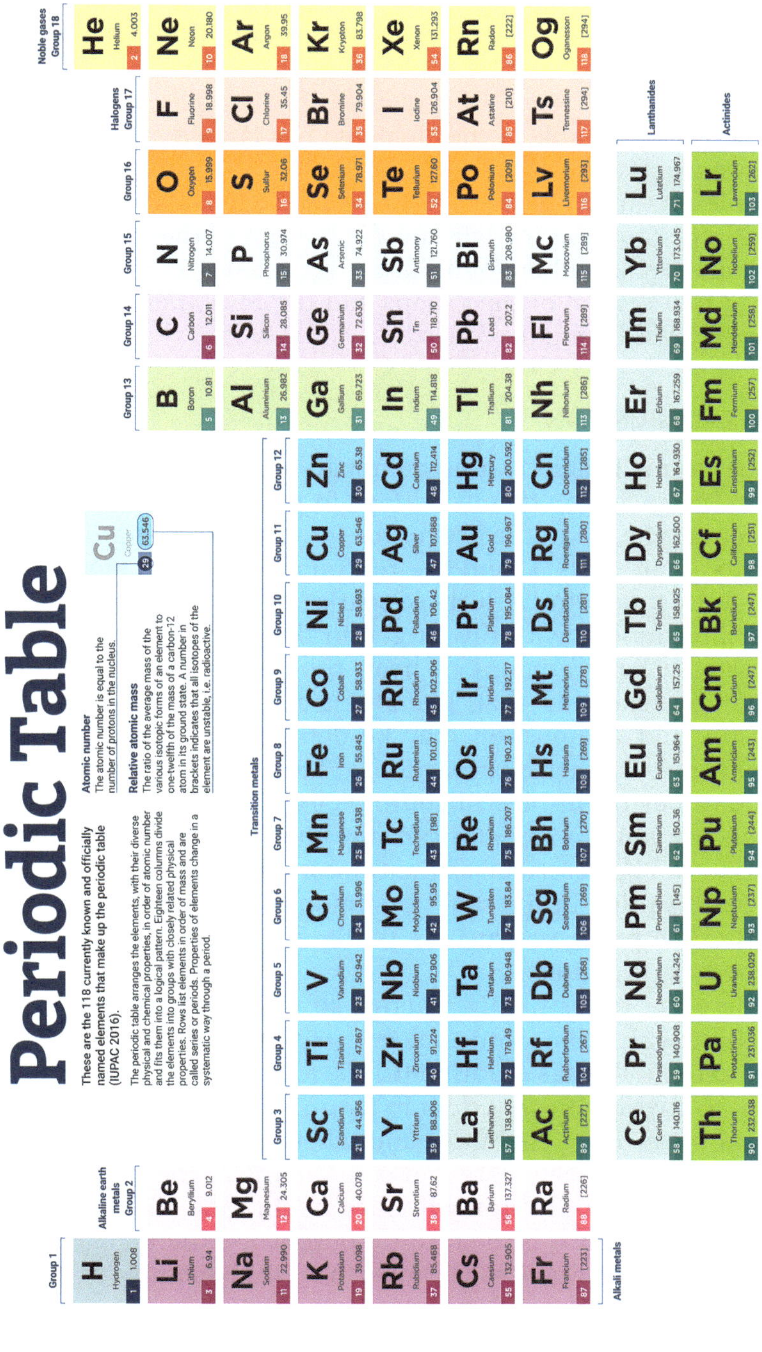
Periodic Table
These are the 118 currently known and officially named elements that make up the periodic table (IUPAC 2016).
The periodic table arranges the elements, with their diverse physical and chemical properties, in order of atomic number and fits them into a logical pattern. Eighteen columns divide the elements into groups with closely related physical properties. Rows list elements in order of mass and are called series or periods. Properties of elements change in a systematic way through a period.
Atomic number
The atomic number is equal to the number of protons in the nucleus.
Relative atomic mass
The ratio of the average mass of the various isotopic forms of an element to one-twelfth of the mass of a carbon-12 atom in its ground state. A number in brackets indicates that all isotopes of the element are unstable, i.e. radioactive.
Alkali metals
Alkaline earth metals
Transition metals
Halogens
Noble gases
Lanthanides
Actinides

Figure 1.1 The Periodic Table.

increases by one. The atoms finally get so large that they become unstable and radioactive. The largest atom in this version of the table belongs to the radioactive element oganesson (Og) with atomic number 118.

The Periodic Table of the elements has an odd shape and this is because as atoms become larger, their structure changes in a complex manner. It is nevertheless a useful arrangement. Metals tend to be found at the left and the bottom of the table. Most of the non-metals will be found in the right hand groups. It will be found that several of the groups and the rows (periods) contain elements with similar properties.

1.2 COMBUSTION

When iron is heated to a high enough temperature in air, it can be made to burn. In our school forge, a rod of iron would sometimes be left in the fire for too long. The temperature would rise to about 1000 °C and the rod once withdrawn would burn brilliantly producing a flurry of sparks. At this temperature, iron combines rapidly with the oxygen in the air and in the process liberates a large amount of heat, producing iron oxide (FeO). Chemical reactions that liberate heat are described as exothermic and combustion may be defined as a rapid and self-supporting exothermic reaction. This means that once the reaction has begun, it will continue as long as the starting materials are able to react with each other. More generally, combustion can be described as the combination of a substance with oxygen resulting in the rapid evolution of heat, light and flame. Exothermic processes such as the example above are often described as oxidations.

There are other highly reactive substances like oxygen that combine with atoms and molecules to produce heat and light. For example the gas chlorine (Cl) is so reactive that it does not occur to any significant extent in nature. Once manufactured it is stored in an inert container and when released it will quickly bond with many elements and compounds in the environment. Some metals such as iron will burn in chlorine and in this case iron chloride, $FeCl_2$ is the result. Combustion therefore does not always involve oxygen so it cannot be defined simply as an exothermic reaction involving oxygen.

The term *fire* has been defined as a *high temperature self-sustaining oxidation* which perhaps surprisingly, does not mention the emission of light which is almost always in evidence. There are many familiar examples of combustion, such as the striking of a match, the burning of a candle and the explosion of petrol in an internal combustion engine. In all of these examples, oxygen combines with one or more elements to produce oxides, liberating heat in the process. In the case of a candle, oxygen combines with the carbon (C) and hydrogen (H) in the wax to yield the oxidation products carbon dioxide, (CO_2) and water (H_2O). In the case of a struck match the products are more numerous. Here the oxygen also combines with elements that include sulphur (S) to yield sulphur dioxide, SO_2. In this case not all of the oxygen comes from the air. Some is present in oxygen-rich compounds in the match head itself. In all of these examples, the bonding of oxygen with another element liberates heat. In a match head, some of the heat is transferred to the wooden stick, raising its temperature until it too takes fire.

Materials that combust are frequently termed 'flammable' or 'inflammable' but the distinction between these two terms is not clear-cut and has occasionally led to confusion. The former term is preferable, allowing incombustible material to be described alternatively as 'non-flammable'.

A certain amount of heat needs to be supplied to most fuels for them to burn in air. Even a match head needs to be primed by an input of heat. This is provided by friction during the striking of the match. If the wooden matchstick without the head had been struck no combustion would result and this shows that different fuels require different amounts of heat energy to combust. The lowest temperature at which a particular fuel will burn in air is called the *auto-ignition temperature*. It varies over a wide range. For the common marsh gas, methane (CH_4), it is 632 °C, a temperature at which iron glows red-hot. For the industrial gas acetylene (C_2H_2) the auto-ignition temperature is only 305 °C. Once ignition has occurred, combustion tends to be self-supporting until either fuel or oxygen runs out.

Spontaneous combustion was first clearly defined by David Frank-Kamenetskii[1] as the combustion of a material in the absence of 'forced ignition', in other words without the employment of a spark or flame to prime the reaction.

Frank-Kamenetskii was born in Vilnius, Lithuania but soon moved east, graduating from the Siberian Technological Institute in 1931. After working in the gold mines of eastern Siberia he joined the Institute of Chemical Physics in Leningrad where he worked on chemical chain reactions. He wrote an influential work on physical chemistry and combustion in 1947.

Spontaneously combusting materials usually have low auto-ignition temperatures and examples will be described in the following chapters.

1.3   FLAMES

A flame is defined as a glowing mass of gas produced during combustion. Several types are recognised. They include 'explosion flames' and 'stationary flames' for example.[2] Explosion flames are those that move out rapidly from a small region in all directions often causing devastation. They are typically of short duration, lasting much less than a second. Stationary flames are more or less fixed in their position and are usually of greater duration- of the order of seconds to days or even years. Stationary flames may be further divided into diffusion flames and pre-mixed flames. In diffusion flames the fuel burns as it contacts the oxygen in the air. The nature of the flame is determined by the manner in which the molecules of fuel mix with the air. If the fuel is a gas such as hydrogen, the molecules of oxygen and hydrogen intermingle prior to combustion. This mixing which is termed diffusion is gradual and it allows the two kinds of molecules to get close enough to react and produce fire without explosion. Diffusion flames have been well studied as they are among the simplest to understand. A good example is provided by the wax candle. Larger flames such as those of a garden bonfire are more complex since they involve stronger mixing of fuel and air. This is the result of large-scale movements of air along with its combustion products in a process known as turbulence. Although diffusion flames appear to result from the simple admixture of fuel plus oxygen, the chemical processes leading to the final products are in fact complex and involve a range of intermediate compounds that usually include free radicals. These consist of atoms, or, more often, molecules with a special electronic feature known as an unpaired electron.

(a)

(b)

luminous zone ⎯⎯⎯

dark zone ⎯⎯⎯

blue zone ⎯
non-luminous zone ⎯

$$C_{22}H_{46}$$

(c)

H-C-C-C-C-C-C-C-C-C-C-C-C-C-C-C-C-C-C-C-C-C-C-H

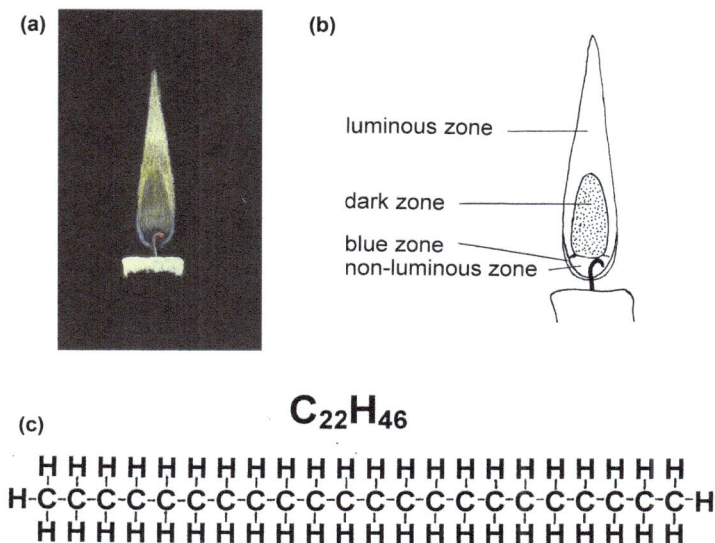

Figure 1.2 (a) Diagram of a burning candle. (b) The candle zones (see the text), (c) the structural and molecular formulae of a molecule of docosane, one of the main constituents of candle wax.

Unpaired electrons have a strong tendency to 'pair' by seeking out weakly-bound electrons in other molecules. This makes free radicals short-lived and extremely reactive.

In diffusion flames the rate of reaction between fuel and oxygen is not considered important. Rather, it is the rate of intermingling that determines its vigour. Flame shape however is determined by mass air movements near the flame. Air close to a flame is warmed and rises because warm air is lighter than cold, a process known as convection. As the warm air pushes up against the flame, the flame is drawn up into its characteristic lanceolate shape. Observations on candles in zero-gravity experiments show that flames assume a spherical shape and become unstable because convection cannot efficiently operate under these conditions.

Candles provide a good example illustrating some important features of combustion (Figure 1.2a and b). Modern candles are composed almost entirely of paraffin wax, a mixture of hydrocarbons obtained from coal or oil by distillation. For candles, the composition of the wax is important because its melting and boiling point must allow it to be drawn up into the combustion zone with a wick. This ensures a stable and long-lasting flame. The wax consists of chains of between 20 and 40 carbon atoms. The mixture melts at a temperature between 45–70 °C. When the wick is lit with a match the heat generated melts the wax immediately below the wick by radiation ensuring a continuous supply of fuel. This is achieved by the wax being drawn up into the fibres of the wick by a process known as capillarity. Modern wicks are of braided cotton that bends over in a flame so that the tip reaches a hot, oxygen-rich zone where it burns away to ash. A bent wick gives a better and safer flame as it provides a larger area for the fuel to evaporate and does not detach to cause potential havoc with nearby flammables.

The normal candle flame consists of several regions, some of which are small and not easily observed. Below but within the bright yellow luminous zone and surrounding the wick is a cone-shaped dark zone where the fuel is vaporised in the absence of oxygen at a temperature of about 1000 °C. In this region the breakdown of large wax molecules (Figure 1.2c) into smaller fuel molecules occurs, by a process known as pyrolysis. Surrounding the dark and luminous zones is a thin sheath of pale blue-coloured flame called the blue- or reaction zone. This is where oxygen from the air intermingles with the hydrocarbon fuel and combusts. Temperatures up to 1400 °C have been recorded here, sufficient to melt a thin gold wire. The pale blue flame consists of light emitted from several processes. These include emissions from free radicals such as CH and O. It is also likely that emission from Swan bands occurs. The light seen in emission bands has a particular range of wavelengths giving it a characteristic colour. The Swan bands were found to be produced by diatomic carbon, C_2 a short-lived molecule formed during the decomposition of wax. William Swan, after whom the bands were named, was a Scottish physicist who discovered them in a wide range of burning hydrocarbon fuels. He concluded that they must be connected to the chemistry of carbon or hydrogen. It was later discovered that the dicarbon molecule (C_2) was responsible using an analytical technique known as spectroscopy.

Moving away from the reaction zone into the bright luminous body of the candle flame the temperature falls to about 1200 °C. Here there is little combustion since the oxygen molecules by this time have almost all reacted with the fuel in the outer re-action zone. Pyrolysis of the wax continues but the temperature is insufficient to turn carbon into a gas and it forms instead small aggregates, better known as soot. These particles glow brilliantly in the heat, a phenomenon known as incandescence. It is this process that allows a candle flame to function as a light source. When the soot particles eventually reach the reaction zone *via* convection they are oxidised and turned into carbon dioxide by the oxygen of the air. However, if the luminous zone is cooled, oxidation is incomplete and the candle smokes with the soot. This can be achieved by playing the flame against a cold non-flammable surface.

The candle flame illustrates well the interaction of oxygen with a fuel and the role played by convection, radiation and diffusion. In the mid-19th century the English scientist Michael Faraday[3] undertook a detailed study of the candle flame. He demonstrated the significance of convection by creating a shadow of the flame against a white screen. He placed a lighted candle in front of the sun and the flame produced a shadow on the screen allowing the convecting gases to be clearly seen.

Premixed flames differ from diffusion flames and are formed when fuel and air are mixed prior to combustion. They have many useful applications and a well-known example is the Bunsen flame. Here fuel is premixed with air at the base of a tube that is then lit at the top about 15 cm distant. These flames are more complex than diffusion flames and they are rarely observed in nature.

Figure 1.3   Incandescence from a hot wire held in a gas flame.

1.4   INCANDESCENCE – THE EMISSION OF LIGHT FROM HOT BODIES

The Latin word *incandescere*, meaning 'to become white' was used by James Hutton to coin a new English word in the late 18th century – *incandescent*. The word was in common use by the mid- and late 20th century to describe the most popular type of electric lighting, produced by a white-hot filament of tungsten heated by an electric current. The key feature of incandescence is light resulting from heat. Most solids and liquids begin to glow a dull red colour at about 520 °C, a temperature known as the Draper Point. The Lancastrian photographer, John W. Draper published observations on incandescence in 1847 while tenured at New York University. It was found that as the temperature of a body increases, the colour of the incandescence changes first to orange, then yellow and finally blue. Research into the relationship between heat and light in the latter part of the 19th century led to the establishment of an important relationship. This is known as the Stefan–Boltzmann Law where the temperature of a perfectly radiating body is related to the radiation that is emitted. Among other things, it helped to establish the temperature of the sun's surface which is about 5500 °C.

Incandescence results from the extreme agitation of atoms and molecules at a high temperature. It is a form of thermal excitation resulting in the emission of radiation (Figure 1.3). As temperature increases, so does the average radiation energy, leading to radiation being emitted at shorter and shorter wavelengths. This results in the change in colour of the light from red to blue. Gases may behave differently from solids and liquids when heated as will be described later.

1.5   RADIATION

Most of the radiation we experience is electromagnetic in nature and consists of oscillating electric and magnetic fields. The radiation is emitted into space as waves and includes X-rays, ultraviolet rays, light, infrared and radio waves. Sources of the radiation include hot bodies or bodies containing electronically excited atoms, molecules or ions. The term radiation has also been extended to include some fast-moving particles, many originating from radioactive sources. Examples are helium nuclei known as α-(alpha) radiation and β-(beta) radiation, the latter consisting of fast-moving electrons. Unless otherwise stated, the term radiation will cover only electromagnetic phenomena here.

Electromagnetic waves are uniquely defined by their ability to pass through space at the speed of light. They can be absorbed by matter which, in the case of light, give materials their characteristic colour. The absorption and re-emission of radiation by matter plays a central role in many of the phenomena described in this book. It is also a subject of extensive research over many years although there still remain areas of uncertainty. This results from the complexity of the processes involved and the challenge of investigating extremely small time scales and events. Time scales shorter than a millionth of a second are often involved. A brief account of radiation relevant to luminous phenomena is given below. This is played out with reference to some of the major discoveries that have helped scientists understand how light interacts with matter.

In the early 1800s Henry Wollaston[4] set up a flint glass prism in a darkened room and allowed sunlight to pass through it using a slit about 1 mm wide. Looking through the prism towards the bright slit he saw the familiar rainbow colours of the light spectrum. He was surprised to see that they were broken up in places by fuzzy dark lines. Wollaston was born in Dereham, Norfolk and studied at Caius College, Cambridge. After working as a physician he became deeply interested in problems of chemistry and metal refinery. In the process he discovered two new elements, palladium and rhodium, and was the first person to find the atomic weight of carbon.

A few years later the German optician Joseph Fraunhofer using superior equipment found over 500 dark lines in the solar spectrum. They remained unexplained until Gustav Kirchoff in 1859 discovered that they were atomic absorption lines. In other words, atoms situated somewhere between the sun's surface and his prism were absorbing light of particular colours (energies). Once it became possible to measure the wavelength of the radiation associated with the lines, it was discovered that there was a simple relationship between the wavelength of the radiation and its energy. Experiments liberating metallic silver from its salts at the birth of photography showed that ultraviolet rays caused a stronger chemical reaction than blue light. Red light had the least effect. It appeared that the shorter the wavelength, and the higher the frequency of the radiation, the greater the energy. Putting these findings together, it was clear that radiation can be absorbed and emitted by atoms at particular energies corresponding to the position of the dark bands that Wollaston and Fraunhofer had seen. In the case of sunlight, most of the absorption bands were found to originate in the sun's atmosphere. They were due to the presence of particular atoms and ions absorbing radiation originating from the deeper layers of the sun.

The nature of light itself proved to be among the most difficult of the physical phenomena to explain. Originally regarded as particulate, it was deemed wave-like by the mid-19th century. The situation changed again when it was discovered that free electrons were emitted by metals when light shone upon them, a process now known as the photoelectric effect. Aided by theoretical work at the end of the 19th century the particle nature of light was again seriously considered. Experimental work such as that undertaken by the American physicist Arthur H. Compton showed that the light particles could interact with free electrons, scattering them in different directions, a process known as the Compton Effect.

The best explanation of theory and experiment saw light as consisting of indivisible energy packets or 'quanta'. The energy of the radiation was found proportional to the frequency of the radiation, the factor of proportion being a small number known as Planck's constant. The term photon was coined in 1916 and is used today to denote a packet or quantum of light energy. It is usually written as the symbol $h\nu$ with h being the Planck's constant and ν the frequency of the radiation in cycles per second. The product is expressed in energy units of electron volts (eV). Radiation is also defined by its wavelength. The distance between the crests of the electromagnetic waves varies enormously within the electromagnetic spectrum but in the visible light region the waves are small and are measured in nanometres (10^{-9} m) (see Box 1.1).

Radiation can therefore be defined by its energy, frequently measured in electron volts; by its wavelength measured in nanometres or by its frequency measured in kilohertz (see Box 1.1). These measurements are interchangeable but their use depends upon the context.

BOX 1.1 THE ELECTROMAGNETIC SPECTRUM

The electromagnetic spectrum is shown in Figure 1.4. It runs from the shortwave gamma rays to the longwave radio waves. Visible light forms a narrow band in this spectrum and includes the peak radiation emitted by the sun. Radiation frequency is the number of waves passing a point per second. It is measured in kilohertz (kHz). One kilohertz represents 1000 cycles per second.

Scientists are often faced with calculations involving exceptionally large and small numbers. Figure 1.4 is a good example. Here a shorthand method has been used to represent the small numbers in the left column of the figure. For example, in the visible region of the spectrum the wavelength of the radiation is not too far removed from a millionth of a metre. This could have been written as 1/1 000 000 or as 0.000001m. Neither is convenient and for the decimal

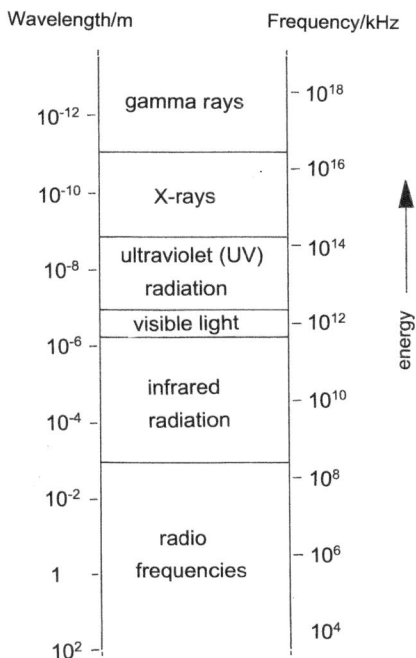

Figure 1.4 The electromagnetic spectrum.

notation mistakes are easily made. Instead an exponential form of the number is used. One million is equivalent to multiplying ten by itself six times and is written as 10^6. A negative sign is placed before the six to indicate 1/1 000 000, namely 10^{-6}. The numbers in the right hand column are large so a negative sign is not shown.

Some of these numbers can be conveniently described in words. Frequently used examples are mega (10^6), kilo (10^3), milli (10^{-3}), micro (10^{-6}), nano (10^{-9}) and pico (10^{-12}).

Photons and electrons, both of which exhibit wavelike- and particle-like properties are largely responsible for the transmission of light and its interaction with matter. One of the best ways of describing the radiation emitted by a body is to plot its intensity against its wavelength as a graph. Figure 1.5 shows the radiation emitted by the sun before it enters earth's atmosphere and by an incandescent lamp. Here you can see that the radiation from the sun reaches a maximum in the green part of the spectrum. Irregularities in the curve are caused by some

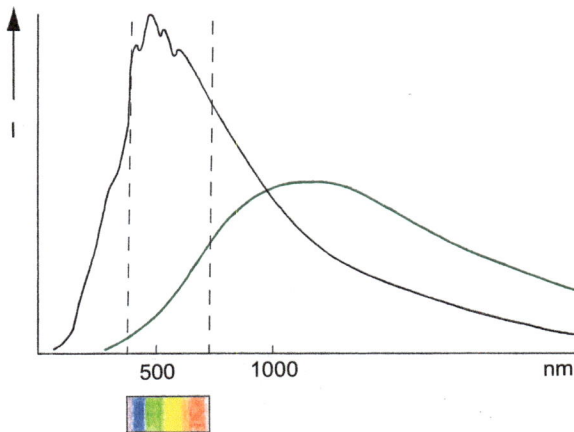

Figure 1.5   Emission of radiation by the sun outside earth's atmosphere (black line) and the hot filament of an electric light (green line). The incandescence spectrum is measured as its wavelength in nanometres and the vertical bar (the ordinate) shows the relative intensity of the radiation (I). Broken vertical lines indicate the range of visible light.

absorption in the sun's own atmosphere. Incandescent lamps operate at a lower temperature than the sun's surface which means that the maximum emission lies in the invisible infrared region of the spectrum. Incandescent lamps have a low content of blue light compared with the sun, and appear yellowish orange when viewed against it. Also note the significant amount of ultraviolet radiation from the sun, to the left of the vertical dashed line. Most of this radiation is absorbed by the earth's atmosphere.

When light is absorbed by an isolated atom, only photons of a particular energy participate. To understand the process in a little more detail it is necessary to delve into the structure of the atom and examine how this absorption takes place. Hydrogen, the simplest atom, consists of two charged particles, a positively charged proton sitting at the centre of the atom and a negatively charged electron orbiting around it. The attribution 'positive' and 'negative' is purely conventional. Positive charges attract negative charges and *vice versa*. All protons are deemed positive and all electrons negative. It is the small electron that is responsible for virtually all chemical processes. The electron's mass is much less than that of the proton and it has more mobility. The centre of an atom is called the nucleus and in the case of hydrogen the nucleus contains just a single proton. But atoms of the other elements are larger. They contain several protons and an uncharged particle called the neutron. The relationship between an atom's electrons and its nucleus has taxed the minds of scientists for over a century. But our understanding of the laws governing these particles is now good. In many cases it allows reliable predictions to be made regarding their behaviour.

The lighter electron was originally thought to encircle the hydrogen nucleus as the earth encircles the sun. This was soon found to be over-simplistic and the electron is now seen as taking up a range of positions around the nucleus, each one of which forms a characteristic three-dimensional pattern. Despite this apparently irregular behaviour, electrons were found to occupy, on average, particular orbits. The orbit closest to the nucleus is spherical but some orbits further out from the nucleus have other shapes. The orbits occupy shells with shell 1 closest to the nucleus. Shell 1 can accommodate up to two electrons, when it is described as full. In this shell the electrons are not identical as they must differ in spin for occupation. As the shell number increases, the energy of the electron increases as does the number of electrons that can be accommodated. Shell 2 for example can accommodate a maximum of eight electrons. The larger shells are conveniently split into sub-shells where the electrons occupy different patterns of distribution. In shell 2 for example there are two subshells denoted 2s and 2p. Electrons in the 2s subshell occupy on average a spherical orbit around the nucleus but in the 2p shell the orbits are more complicated. The electrons also have slightly different energies in these subshells. Shell 2 therefore is more complex than Shell 1 and in both shells, a maximum of two electrons can occupy each individual orbit. Hydrogen is a simple case as it has a single spherical subshell denoted 1s as it has only one electron.

Electrons can jump from one shell to another only if they gain or lose energy. This can be effected by radiation. In the case of the hydrogen atom, red light of energy 1.89 eV can cause an electron to jump from shell 2 to shell 3 so that energy is transferred from the radiation to the orbiting electron. Transfers can also occur from shell 2 to other shells producing a series of energy changes that increase in size as the jumps become larger. When an electron falls back down to shell 2 from a higher shell, energy is lost. For hydrogen atoms, the energy is carried away by a photon of matching energy. Since energy is neither created nor destroyed, all of the photon's energy is absorbed by the atom, increasing its energy by the same amount.

Emissions of radiation from hydrogen are common in stars and nebulae and the red emission line of hydrogen is prominent in the Orion nebula. This is caused by the descent of an electron from shell 3 to shell 2. An electron promoted from shell 2 to shell 4 requires absorption of a photon in the more energetic blue region of the spectrum. Climbs to even higher levels are only possible with radiation more energetic than that provided by light, such as that in the ultraviolet region. Transitions involving shell 2 for the hydrogen atom, belong to the Balmer series and are illustrated in Figure 1.6. Transitions from levels 3 to 2 and from 4 to 2 are in the visible region. When an electron falls in energy from 3 to 2 a photon of red light is emitted as shown by the red bar in the figure. The fall from 4 to 2 is of higher energy and blue light is emitted. More energetic transitions such as

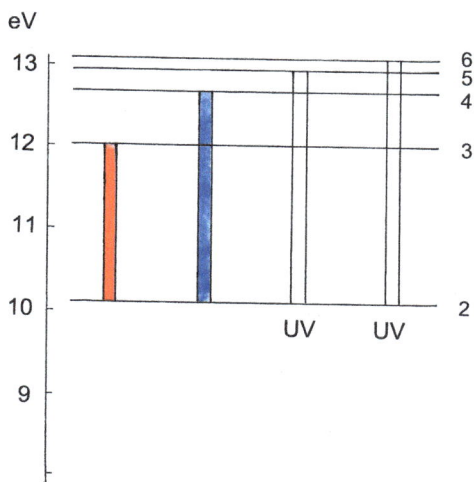

Figure 1.6 The Balmer series for the hydrogen atom. Energies in electron volts are shown on the left of the figure.

5 to 2 and 6 to 2 result in photons in the ultraviolet region of the spectrum and are invisible. The series is named after the Swiss mathematician Johann Balmer who discovered it. There are other named series covering transitions to shell 1 and to shell 3 but none of these emissions and absorptions are in the visible part of the spectrum.

For larger atoms and molecules, the electron transitions are more complicated. In the case of larger atoms this is because they contain more than one electron. The charges on these electrons cause their movements to be modified as like charges repel. An example is the noble gas neon, sometimes used to colour advertising lights. If a high voltage electric current is passed through this gas the electrons are excited to a wide range of energy levels leading to the formation of a large number of emission lines. In the visible range, most of these lines are in the red region of the spectrum so the gas glows with a bright red light (Figure 1.7).

The situation in molecules is further complicated by the presence of more than one nucleus in the vicinity of the electrons. Being small, electrons are influenced by nuclei of other nearby atoms and this can lead to an electron being shared between atoms. This result is the bonding of atoms to form molecules, the strength of the bond being itself influenced by many

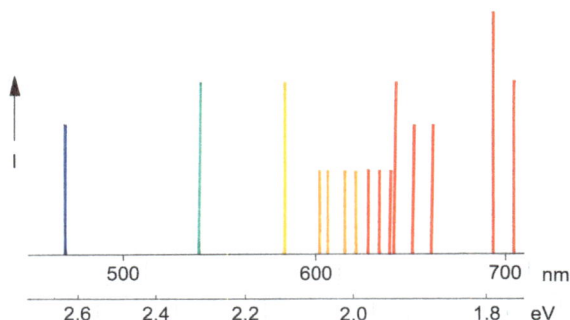

Figure 1.7   The atomic emission lines of the noble gas neon in the visible region of the spectrum. Lines are plotted against their wavelength and their photon energy in electron volts. Approximate intensity (I) is plotted on the ordinate.

factors. So although photons are absorbed by molecules in essentially the same way as in the hydrogen atom, the transitions to higher or lower energy levels by the electrons are modified and are much more numerous. Some of the more important modifications are caused by the vibrations and rotations of the nuclei, both of which take on specific energies. With these additional complications, the emission and absorption spectra of collections of molecules are usually more complex than those of atoms. This often results in broad emission bands rather than sharp spikes. Additional factors often include effects of impurities and imperfections within the molecular/crystal structure which may give rise to phosphorescence (see below).

Some of the energy levels for an atom/molecule are shown in Figure 1.8a. Here the horizontal lines represent the energy values which an electron can take with energy increasing from the bottom to the top of the page. The thick lines represent the lowest energy level of a particular electron shell and are lettered S_0, S_1, S_2, *etc*. The lowest level is known as the ground state and is the level where the electron is usually to be found. Within each shell a series of thinner lines can be seen. These represent vibrational energy levels. They are sometimes labelled v_1, v_2, v_3... of increasing energy. Each shell has a large number of these levels and only a small number are shown in the diagram. Note how the difference between these levels becomes progressively narrower for both the shells and the vibrations as the energy

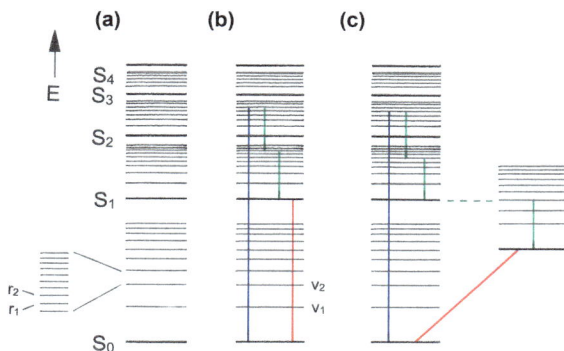

Figure 1.8   Energy levels, fluorescence and phosphorescence. (a) A series of
energy levels shown by horizontal lines. Thinner lines represent
vibrational energy levels (v_1, v_2). Finer rotational divisions, (r_1, r_2
etc.) are shown to the left. (b) Fluorescence. An electron is
promoted to a higher energy level (blue vertical line). On returning to the ground state, some energy is emitted as heat
(green lines) and a photon of a lower energy (red line). (c)
Phosphorescence. The electron, upon reaching S_1 undergoes a
change in spin state before a photon is emitted.

increases. The region between v_2 and v_3 in shell S_0 has been
expanded to the left of the figure to show even finer divisions
resulting from quantised rotations as r_1, r_2, r_3 and so on.

We have seen that when a photon is absorbed by an atom or
molecule an electron is promoted to a higher energy level. It is
then said to exist in an excited state. This simply means that the
energy of the electron has been increased above the ground or
stable state. This is illustrated in Figure 1.8b. Here a photon has
been absorbed by a molecule and the energy transferred to an
electron. The electron has been promoted to a higher more energetic state, as shown by the blue line. In this case the photon's
energy has transported the electron to one of the lower vibrational levels of shell S_2. Excited states tend to be short-lived.
Usually the electron soon returns to its former ground state,
losing energy as it does so, in agreement with the law of conservation of energy.

In the ground state, the electrons of most atoms and molecules exist at the lowest energy level that they encounter
among the different shells. This state is however temperature
dependent. As temperature increases, the more violent

agitation of the atoms can promote electrons to one or more excited states.

1.6   LUMINESCENCE

Most of the light we see in the natural world is incandescent and comes from the sun, but there is a small proportion originating from other sources, collected under the umbrella term *luminescence*. In contrast to incandescence, luminescence is the emission of light from bodies unconnected with a rise in temperature. Instead it is the result of an interaction between an atom or a molecule and an energetic photon or a particle without the participation of heat. The particle/photon may originate from an external energy source or from within the material itself. Most luminescence is the result of an interaction producing an electronically excited electron. When the electron returns to a ground state a photon is emitted. It is important not to confuse luminescence with *luminous*. The latter term is in more general use and includes luminescence along with incandescence and any other form of light reflected or refracted from a body.

There are several kinds of luminescence and most of them are known by their prefixes. Some of these terms address the effects rather than the causes of light emission as will be seen below.

1.6.1   Photoluminescence – Fluorescence and Phosphorescence

Photoluminescence is light emission from an object that results from the absorption of photons. As we have already noted, absorption causes some of the electrons to move into higher energy levels. Although the term implies that the exciting radiation is confined to the visible range, the effect of more energetic but non-visible radiation such as ultraviolet radiation (UV) is often included.

In the early 1800s the German chemist Johann Ritter experimented with sunlight passed through prisms and noticed that silver chloride (AgCl) was rapidly converted into silver metal. This happened when the prism was adjusted just beyond the violet end of the spectrum and played upon the salt. The cause was attributed to 'chemical rays' later to be known as ultraviolet radiation. David Brewster, the 'Father of modern experimental optics' had also discovered the phenomenon in the mineral fluorspar.[5] Brewster (1781–1868) was admitted to the University of Edinburgh aged 12 and later became a minister of the Church of Scotland. After graduating he soon became interested in physics and liaised with several well-known scientists of his day. He became fascinated by the passage of light through crystals and published widely on the topic. He is also known as the founder of the British Science Association and the inventor of the kaleidoscope. Brewster found that when fluorspar was placed in the 'chemical rays', light was emitted along some of the crystal planes. He suspected that it was due to some substance taken up during the crystallisation of the mineral.

Later, the Irish physicist George Stokes carried out similar studies using prisms and played the sun's rays on a range of solids and solutions.[6] He discovered that several of these materials became illuminated by the invisible rays. Among the substances tested, green fluorspar, a decoction of horse chestnut bark and Canary glass were particularly effective. Horse chestnut bark contains a substance known as aesculin and this was found responsible for light emission (Figure 1.9) while Canary glass contains an emitting uranium salt (Figure 1.10). Stokes also discovered that the wavelength of radiation emitted was always longer than the wavelength of the radiation absorbed. He was the first person to use the term fluorescence after observing the emission in fluorspar. The difference in wavelength between the absorbed and emitted radiation later became known as the

Figure 1.9 Sections of twig from the horse chestnut, *Aesculus hippocastanum* illuminated with UV radiation and showing fluorescence due to aesculin under the bark.

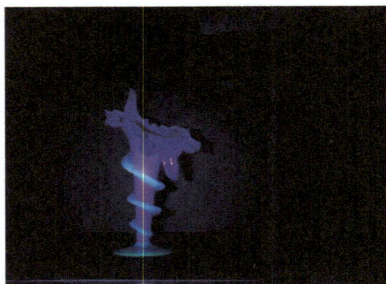

Figure 1.10   Fluorescence of canary glass illuminated by UV radiation.

Stokes shift (Figure 1.11). This occurs because some of the energy absorbed by the atoms is lost through other processes such as heat before it is re-emitted.

So some of the energy gained by the electrons in solids and liquids is dissipated as heat resulting in the increased vibration of the atoms. The heat is then transferred back into the environment. Stepwise losses take place with many small energy drops composed of minute fractions of an electron volt until a favourable energy level is reached for a photon to be emitted.

Illustrated in Figure 1.8b is the process of fluorescence. Here an excited singlet spin state electron (defined below) loses its energy in steps shown by the green and red lines in the diagram. It first loses energy by falling into one of the vibrational states of a lower shell, in this case S_1. This type of transfer is only possible if the vibrational energy levels of the two shells become coupled. Coupling can only happen when the energy levels of the two shells are close together, as seen near the top of the diagram. When an excited electron loses energy in falling to a lower excited state the process is known as internal conversion. No light is emitted and the energy is lost as heat. Within shell S_1, energy is further lost as heat when the electron falls to the lowest vibrational level shown by the lower green line. This process occurs only within a particular shell and is known as vibrational relaxation. Again no light is emitted. In the final stage the electron falls back to the ground state and a photon carries away the energy (red line). It is this light that is seen as fluorescence. This final stage is fast compared with the other two. It happens in about a nanosecond, $(10^{-9}$ s) – equivalent to the time it takes a fired rifle bullet to travel a thousandth of a millimetre.

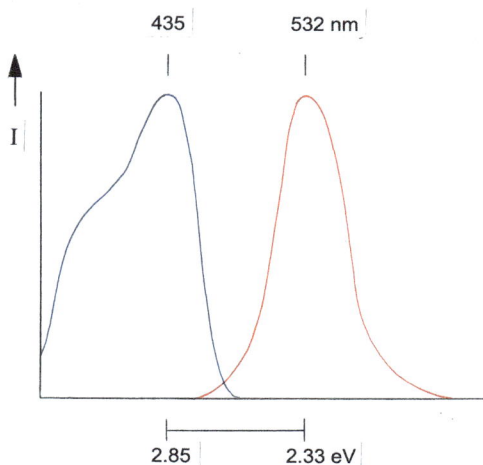

Figure 1.11 Illustration of a Stokes shift for an organic compound irradiated with UV with a maximum intensity of 435 nm (blue line) with fluorescence of a maximum intensity 532 nm (red line). Note that energy decreases and wavelength increases from left to right in this figure. Intensities normalised. The Stokes shift shows an energy change of 0.52 eV from the ultraviolet maximum (2.85 eV) to the visible maximum (2.33 eV).

One of the more interesting ways of examining fluorescence is to observe the diffusion of antimony chloride into potassium chloride. Pure potassium chloride is not fluorescent but if a small quantity is placed on a warm glass slide and a similar amount of antimony chloride is placed next to it without touching, the antimony will volatilise and some of the molecules will diffuse into the potassium salt. As they do so, the potassium salt gives a bright blue fluorescence under ultraviolet light. Antimony ions diffuse into the crystal structure of the salt, altering its physical properties so that a photon can be emitted. Since antimony is poisonous the experiment must be undertaken in a fume cupboard. Nowadays we are surrounded by fluorescent materials and they are easy to examine with an inexpensive UV torch. The paper of this book contains a fluorescent brightener and most washing powders also contain them. Brightneners include the aminated sulphonic acids. They show a strong blue-white fluorescence caused by the UV in sunlight. They contain amine (NH_2) and sulphonate (SO_3H) groups attached to aromatic hydrocarbon rings.

In atoms, electron pairs with opposite spin are termed singlets and they are said to exist in a singlet spin state. An electron pair with the same spin is termed a triplet state. The Pauli exclusion principle does not allow electrons with the same spin to occupy the same orbital and it has been found that a triplet to singlet transition is slow when compared with other electronic processes.

Phosphorescence shares some features with fluorescence and the two mechanisms are often confused. In Figure 1.8c one of the mechanisms is shown. Initially the same process occurs as in fluorescence. In the diagram a singlet state electron is promoted to an excited state by the absorption of a photon. When the excited electron reaches the lowest state in S1 *via* internal conversion it undergoes another kind of change. The excited singlet state electron changes to an excited triplet state *via* a process known as inter-system crossing (broken green line in Figure 1.8c). The excited triplet then falls by internal conversion to a lower energy and finally returns to the ground state with emission of a photon. Intersystem crossing is an electronically slow process but the time difference is usually too short to be picked up by the human eye.

There is a second phosphorescence pathway involving electron trapping. Crystals frequently contain imperfections. These can be structural defects or the inclusion of foreign atoms or molecules. Electrons, excited by an absorbed photon, can become caught in these imperfections and remain in an excited state for periods of minutes or even years in regions called traps. They require a spike of energy to release them and a photon is then emitted through the processes described above. The energy spike is believed in most cases to be caused by random thermal vibrations. So-called deep traps cause strong electron binding. In these traps electrons may require an exceptionally large spike of energy to release them. Deep traps are thought responsible for much long-term phosphorescence, the strength of which is related to the density of trapping sites within the crystal. This type of phosphorescence is sometimes called persistent luminescence to distinguish it from short-term phosphorescence. Chemists distinguish phosphorescence from fluorescence using a delay time that is greater than 10^{-8} seconds. When a phosphorescent material is illuminated and the light source extinguished, the material will continue to glow. But when the delay time for emission is short, the human eye cannot distinguish fluorescence from phosphorescence.

1.6.2 A guide to the Types of Luminescence

We now see that light is emitted from materials as the result of several unrelated processes and we have noted several terms for these: incandescence, luminescence fluorescence and phosphorescence. These processes are among the most frequently encountered but others are known. Scientists have attempted to classify these emissions in terms of the mechanisms involved. This has resulted in the adoption of more than 20 luminescence prefixes to describe them. Such a classification can help to explain light phenomena but it should be noted that more than one process can occur during emission. In some cases the emission is not fully understood so the application of these names is sometimes uncertain. The main luminescence prefixes are tabulated below in alphabetic order. They are based upon the principal processes believed to be involved. An explanation of each follows, with the exception of those already described. It will be noted that many terms are associated with charge recombination and that some of them describe basically the same process. Most of these processes will be encountered again in the following chapters.

1.6.2.1 Anodoluminescence. See radioluminescence.

1.6.2.2 Bioluminescence. This is chemiluminescence occurring within or caused by living organisms.

1.6.2.3 Candoluminescence. Not all of the light given out by hot bodies is due to incandescence. There is also a process shown by a small number of chemical compounds that produce far more light than predicted by their temperature. Examples are the burning of magnesium metal in air and the heating of calcium oxide. Both produce an extraordinarily bright light which has been referred to as candoluminescence. The mechanism is thought to be the result of an imbalance giving greater energy release in the visible part of the spectrum balanced by low emission in the thermal region. It probably involves emission due to the recombination of free radicals in the hot gas.

Calcium oxide was first used to illuminate outdoor entertainment on Herne Bay Pier, Kent in 1836 and soon found its way to the stage where it was called limelight. A small block of the oxide was heated to a high temperature using an oxy-hydrogen blowpipe and the light focused onto the actors. It was also used in the American Civil War to illuminate enemy targets. It was inconvenient to use and was replaced with electric lighting by 1900.

1.6.2.4 Cathodoluminescence. This is a form of electroluminescence resulting from fast electrons hitting a 'phosphor' screen to produce light. An example is the cathode ray tube, once incorporated into television sets. A thin layer of phosphor is sprayed with fast electrons accelerated in a strong electric field in a vacuum. Interaction between the electrons and the phosphor produces a bright image on the screen. It is also used to produce images in the electron microscope.

1.6.2.5 Chemiluminescence. When a chemical reaction occurs, the reactants, consisting of one, but usually two or more chemical compounds undergo chemical change to form products. During the process several physical processes occur and in most cases this includes the release of energy. The term was first used in 1877 with the discovery of the chemiluminescent compound known as lophine. The mechanisms of bioluminescence are those of chemiluminescence but confined to living organisms and catalysed by enzymes.

In chemical reactions involving chemiluminescence, a small fraction of the released energy is converted into light. This is because some of it is bound up in excited electrons. These electrons sooner or later return to a ground state in a process similar to fluorescence, emitting a photon in the process. Oxygen molecules are often involved in these reactions and the mechanisms are sometimes called by the alternative name oxyluminescence. One of the best known examples of chemiluminescence in chemistry is the oxidation of white phosphorus. Because this form of phosphorus (symbol P) is also spontaneously combustible in air, it is cast into small sticks and kept under water. If a stick is carefully dried and placed on a piece of fireproof material in cool air it is seen to glow in the dark. The stick slowly vapourises in the air. P_4 molecules present in the vapour then react with the oxygen to form short-lived and unstable oxides in an excited state. De-excitation leads to the emission of light (Figure 1.12a). The emission spectrum is continuous in the visible region but it also contains some discrete emission lines in the UV.[7] Equally interesting is the cold flame of phosphorus. This can be made by placing a few hundred milligrams of white phosphorus in a one-litre flask. The flask is flushed with a supply of carbon dioxide gas and the resulting P_4 vapour is allowed to exit *via* a small tube. At the exit, a chemiluminescent cold flame appears (Figure 1.12b).

Early demonstrations of phosphorus were entertaining for their audiences. White phosphorus, dissolved in a suitable liquid, would be handed around and the audience would proceed to daub their hands and faces then delight in the effect as the room was darkened. The highly poisonous properties of the element were not recognised at the time so this type of entertainment was stopped many years ago. The dangerously radioactive element radium, whose salts also glow in the dark was treated in a similar way with dire consequences.

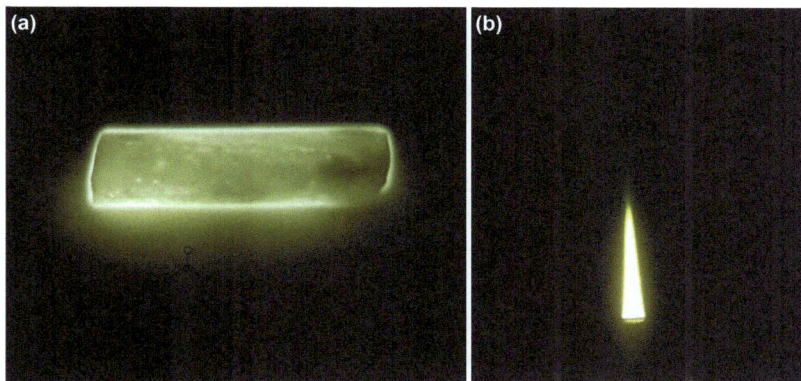

Figure 1.12   Chemiluminescence in white phosphorus. (a) A stick of phosphorus exposed to the air glows as it oxidises at room temperature. (b) The cold flame of phosphorus. The vapour, diluted with carbon dioxide, has been passed through a thin glass tube and chemiluminesces when it contacts air. Flame shape is related to the flow of gas, not convection.

Another interesting example of chemiluminescence is shown by sodium. If a warmed stick of sodium is cut with a knife in the dark, it immediately reacts with the oxygen and water vapour in the air to form sodium hydroxide. It is accompanied by a visible chemiluminescence (Figure 1.13).

The faint orange glow sometimes seen around deep sea vents is thought to be caused by the chemiluminescent reaction between upwelling sulphides and oxygen in the sea water

1.6.2.6 Cryoluminescence. Cryoluminescence is a form of mechanoluminescence. Light emission upon cooling or freezing. There are few examples. The mineral astrakanite (sodium magnesium sulphate) is said to glow upon freezing.

1.6.2.7 Crystalloluminescence. Chemists first reported on light emission by rapidly crystallising substances in the 1800s, initially with potassium sulphate in water. Rapid crystallisation can be achieved by the 'common ion' effect. For example, the swift precipitation of a saturated common salt solution by the addition of strong hydrochloric acid. The acid provides an excess of chloride ions and the salt is immediately formed as small cubic crystals. A bluish light is emitted for a few seconds upon mixing but is not always observed. Studies have shown that an additional ion is required for good emission. Traces of lead have been found to be effective, and this element is also

Figure 1.13 Chemiluminescence of warm sodium metal photographed in the dark. The stick measures 12 × 18 mm.

known to be an activator in alkali halide phosphors. The phenomenon of crystalloluminescence was long suspected as being a form of triboluminescence. The extremely rapid crystallisation, it was argued, could cause friction and fracturing of the crystals during growth. However in the case of the alkali halides spectroscopic measurements and calculations suggest that the light is emitted as the crystal nucleus grows into the crystal lattice (the crystal framework). However, this hypothesis may not be correct in all cases. Crystallisation of barium chlorate for example is accompanied by audible cracking during light emission, a clear sign of triboluminescence. The phenomenon has been little studied.

1.6.2.8 Electroluminescence. Electroluminescence is defined as the emission of light resulting from the passage of an electric current or a strong electric field. Electrons move more slowly through solid material than through space because they interact with matter. Electrons flowing through a copper wire as part of an electric circuit flow at about 1 cm per second. Even at these speeds they can be made into effective emitters of radiation. Free electrons can attain a high speed when accelerated in an electric field. This has been observed mainly in gases since here the molecules are spread far apart. As a result, the electrons can travel greater distances before they encounter them. As electrons collide with neutral atoms, fragmentation often occurs with the production of further free electrons and positive ions. These charged regions are known as plasmas and they are largely responsible for the production of auroras and St Elmo's Fire. They are further discussed in Chapter 5.

Some electroluminescence probably occurs in meteor trails which often show a brief after-glow. Meteors upon entering the atmosphere are subjected to 'ram pressure' causing them to be heated to a high temperature with incandescence. The resulting luminous trail consists of small meteor fragments and ionised gases, including electrons. The ions soon recombine with the emission of light. Spacecraft re-entry trails show the same effect.

Electrically-produced light emission is frequently reserved for electronic devices such as light-emitting diodes (LEDs) and solar cells. The light-emitting diode is currently a major source of artificial light that has superseded most other illumination sources on account of its high energy efficiency and adaptability. Further information on LEDs can be found in Box 1.2.

BOX 1.2   LED OPERATION

Wide bands as opposed to discrete energy levels occur in large collections of atoms as in solids. They consist of large numbers of individual electron energy levels since in solids the energy relationships are more complex. This leads to many combinations of states. In such collections of atoms 'valency bands' and 'conduction bands' can be recognised. The valency bands include the energy levels of the valency electrons – those electrons furthest from the nucleus that are involved in chemical reactions. The conduction bands consist of a group of energy levels that exist above the valency bands and of higher energy still. The bands may be separated by a gap known as the 'forbidden zone' where there are no energy levels for the electrons to occupy. In good conductors such as copper, the valency and conduction bands overlap but in insulators and semiconductors there is a wide energy gap between them (Figure 1.14a–d).

In cold solids electrons always tend to occupy the lowest energy states. This minimises the total energy of the material.

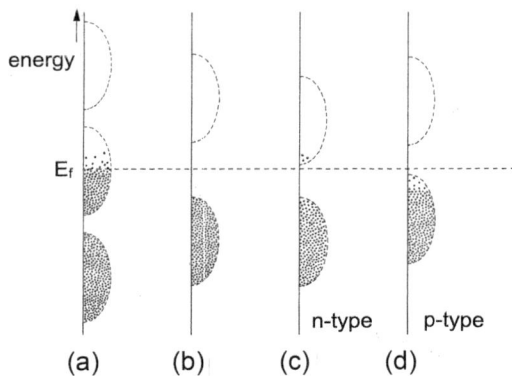

Figure 1.14   Occupation of energy bands in (a) conductors, (b) insulators, (c) n-type semiconductors and (d) p-type semiconductors. Stippling represents electron density.

Since each state can only accommodate one electron the higher energy states will only be filled once the lower levels are full. The states are filled 'bottom up' until a level is finally reached where all electrons occupy states. This level corresponds to what is known as the Fermi energy (E_f in Figure 1.14). The Fermi energy represents a dividing line between the valency electrons of lower energy and the conduction electrons of higher energy.

There are also materials called semiconductors. As their name suggests, they conduct more electricity than insulators but less electricity than conductors. They also differ in conducting electricity in only one direction. Commercial semiconductors consist of insulators such as silicon with added impurities or 'dopants'. The resulting electronic environment allows some electrons to pass through the 'forbidden zone' into the conduction bands. A common dopant is phosphorus. Phosphorus contains one more electron than silicon in its valency band and this electron is able to jump to the conduction band where it can conduct electricity.

In Figure 1.14 electron bands are shown in the vicinity of the Fermi level (E_F). Dense stippling represents bands filled with electrons. In (a) one of the bands crosses E_F showing that there is continuity between a valency and a conduction band. This is the situation in good conductors of electricity. In (b) there is a large gap between the two bands and the energy jump for electrons is too great to allow sufficient numbers to pass across *via* their thermal motions. As a result, conduction of electricity is limited and insulators fall into this category. Figure 1.14c and d represent two types of semiconductors. In these materials, either the valency band or the conduction band lies close to the Fermi level allowing some flow of electricity. Semiconductors that allow a few electrons to enter the conduction band are called n-types, because they carry the negative charges or electrons (Figure 1.14c). Semiconductors where the dopant is an element with fewer valency electrons than silicon such as boron are known as p-types (Figure 1.13d). In this case, the material is deficient in electrons. During the passage of electric current, the atoms with 'missing' electrons become neutralised as electrons pass through the material. Atoms with missing electrons are described as 'holes'.

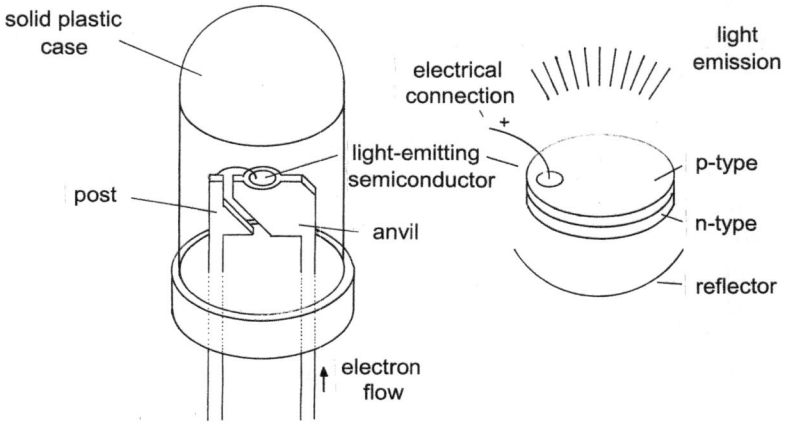

Figure 1.15   Diagram of a light-emitting diode. See text for details.

A simple diagram of an LED is shown in Figure 1.15. Electric current is facilitated by two metallic electrodes known as the anvil and the post. The anvil supports the LED assembly which sits on top. It consists of a thin wafer of gallium arsenide doped on its lower side to make an n-type and on its upper side to make a p-type semiconductor. Electrons flow up the anvil and enter the n-type and pass on to the junction with the p-type. Electrical continuity is maintained with a fine wire. As the holes of the p-type move toward the junction they meet the incoming electrons from the n-type. Here energy is released and light is emitted. Light produced below the semiconductors is collected by a reflector to increase its intensity. The entire structure is enclosed in a plastic housing to protect the electronics, provide insulation and help focus the light. Other designs of LEDs are used in the screens of some modern televisions and mobile phones. OLEDS (organic LEDs) are incorporated into a complex multi-layered structure containing a liquid crystal display providing high contrast and good definition of images.

1.6.2.9 Fractoluminescence. See triboluminescence.

1.6.2.10 Galvanoluminescence. This form of electroluminescence occurs during the process of electrolysis when electrodes are immersed in an electrolyte and an electric current is applied. The positively charged ions, usually metal ions in solution are attracted to the negatively charged cathode and the negative ions move toward the anode. Faint light emission from the anode is sometimes observed and has been termed galvanoluminescence. The emission is associated with thin oxide films that develop upon the electrode and it increases with voltage. Electrolysis using aluminium anodes has long been known to cause light emission. Using a current density of 60 milliamps per square centimetre of the anode is said to give sufficient light to read by. Emission occurs with a range of acids, alkalis and other electrolytes so it is clearly related to the aluminium rather than the electrolyte. It is almost certainly associated with a thin layer of aluminium oxide. Glass electrodes coated with a film of tin oxide using sodium borate or bromide as electrolyte also galvanoluminesce. Emission appears to be associated with a range of negatively charged ions contacting an oxidised anode surface with electrons moving from the ion to an oxide film.

1.6.2.11 Lyoluminescence. Light emission is produced when crystals dissolve. It is rarely encountered, an example being the solution in water of potassium chloride that has been bombarded with cathode rays (fast electrons).

1.6.2.12 Mechanoluminescence. An umbrella term including light emission caused by the deformation or fracture of solids. It has been further divided into triboluminescence, also known as fractoluminescence, (fracturing), plastoluminescence (plastic deformation), piezoluminescence (pressure effects), crystalloluminescence (rapid crystallisation) and cryoluminescence (rapid cooling).

1.6.2.13 Piezoluminescence. Light emission from crystals subjected to pressure, resulting in deformation of the crystal structure. As a result some electrons are released charging the crystal and creating recombination centres in the crystal lattice. Piezoelectricity may also lead to gas discharge in substances such as sugar and contribute towards triboluminescence. It is known to occur in irradiated alkali halides and quartz.

1.6.2.14 Photoluminescence. A general term for light emission following the absorption of photons leading to electron excitation and light emission. Most forms of fluorescence and phosphorescence are classified as photoluminescence. One example of photoluminescence is the laser where energy is supplied to a material known as a gain medium, usually a solid or gas in such a way that the emission is greatly enhanced by a process of repetition. It is also the type of emission found in quantum dots.

1.6.2.15 Pyroluminescence. Light emission from atoms heated to a high temperature. This type of luminescence is limited to gases where particular wavelengths of light are emitted. Examples are the emission of yellow light by sodium and of red light by strontium (Figure 1.16). It is better known as flame-emission and is sometimes included with incandescence, since high temperature is involved (compare luminescence definition above).

1.6.2.16 Plastoluminescence. Light emission *via* plastic deformation. It has been detected when rubber is stretched almost to its breaking point. Thin crystals of zinc sulphide doped with manganese and subjected to plastic deformation also emit light with the manganese acting as a luminescence centre.

Figure 1.16   Strontium flame emission (pyroluminescence) in the red part of the spectrum at around 620 nm.

1.6.2.17 Sonoluminescence. The sound vibrations detectable by our ears range from around 16 Hertz (16 vibrations per second) to 16 kHz. Higher frequencies of sound are known as ultrasound and above about 20 kHz interesting effects are seen in liquids. At 20 kHz the wavelength of the sound waves is around 8 cm, the waves consisting of alternating high and low pressure zones. In the low pressure part of the cycle a liquid is under tension and microscopic bubbles can appear in the vicinity of impurities and gas molecules. The bubbles fill with gas and during the following high pressure part of the cycle the gas is compressed adiabatically resulting in temperatures of around 5000 °C and pressures of 1000 atmospheres (10^8 Pa) or more. This momentary collapse is known as cavitation and during it, interesting chemistry has been observed in the resulting plasma.[8] Research into photographic processes in the 1930s resulted in the discovery of small collapsing bubbles that emitted light. The emission results mainly from incandescence. It is described further in Chapter 7.

1.6.2.18 Thermoluminescence. This is a form of photoluminescence. It applies to rocks and minerals that have been exposed to radiation. Excitations are produced and electrons fall into traps. Heating the material at a later date can lead to light emission. During heating, crystal lattice vibrations known as phonons are produced resulting in the electrons returning to the ground state with emission of light. Excitation can be produced by the products of radioactive decay within the rock or from energetic photons entering the material from outside.

1.6.2.19 Radioluminescence. An umbrella term for light emission caused by ionising radiation formed during radioactive decay – the alpha-, beta- and gamma rays. Gamma rays are energetic photons and beta rays are energetic electrons. Gamma rays could be equally well categorised under photoluminescence since photons are involved. Alpha rays are energetic particles consisting of positively charged helium nuclei. These rays have a much greater mass than electrons and are capable of disrupting a large number of atoms along their path. The result is a shower of diverse atomic fragments capable of causing light emission. The luminescence produced by alpha rays is sometimes referred to as anodoluminescence owing to their positive charge. The artificial isotope promethium 147 produces β-rays (fast electrons) during its decay, and mixed with a suitable phosphor was used to make luminous dials for watches and other instruments. Radioluminescence is also associated with other highly radioactive elements such as radium and the noble gas radon.

1.6.2.20 Triboluminescence. The light emitted through the fracturing of crystals has been known for hundreds of years because it is encountered in a common foodstuff, sugar. Diamond and quartz are other well-known examples. It has been estimated that up to one half of all inorganic compounds are triboluminescent together with a good number of organic compounds. Among all of these, it appears that only those without a centre of symmetry in their crystals show this behaviour. Crystals with a centre of symmetry have opposite and parallel faces of a similar form. Many of these materials emit so little light that it is only detectable using a photomultiplier.

Ingenious studies fracturing crystals in different noble gases have shown that much of the light emission is characteristic of the gas used. As a triboluminescent crystal fractures, the two broken faces can end up with opposite electric charges. Electrical breakdown in the gas occurs across the fracture and leads to a discharge and light emission. Fracturing in air gives spectra characteristic of nitrogen and occasionally water vapour and oxygen. Spectroscopic studies have confirmed this mechanism in several triboluminescent materials but not all.

The excited gas molecules also emit some UV leading to photoluminescence in the fractured minerals. These frequently contain F-centres and other impurities that emit light at other wavelengths detectable spectroscopically. Both fluorescence and phosphorescence have been reported in this photoluminescence. Other light-emitting processes are probably involved during crystal fracturing. High temperatures may occur at the tip of a crack leading to local incandescence. Some crystals also emit light upon the application of pressure resulting from piezoluminescence. Despite intensive research into triboluminescent materials, particularly over the past 20 years, details of these mechanisms remain elusive.

Triboluminescence has many useful applications. A material designed to be embedded in carbon composite structures has received interest in the world of aviation. Bird strikes on planes can be dangerous but are difficult to detect. By coupling a triboluminescent material to an optic fibre, bird collisions could be spotted quickly and the hazard assessed. Research has also been directed into finding triboluminescent materials allowing robots to detect surfaces.

1.7 CERENKOV RADIATION AND BREMSSTRAHLUNG

Not classified as one of the 'scences' and unlikely to be seen unless you work in a nuclear power plant, Cerenkov radiation results from a type of shock wave similar in some ways to the sonic boom of an aircraft but involves light rather than sound. Light travels at lower speeds through matter than it does through space. An electron with a speed close to that of light could, by entering a different medium, exceed that speed and in doing so emit radiation. Most of the radiation is of high frequency and usually visible only at the violet or blue region of the spectrum. It can be seen when highly radioactive materials emitting fast electrons (β-rays) are placed in water. It is also thought to be responsible for 'astronaut's eye'. These are random light flashes resulting from cosmic radiation being caught by the eye during space travel.

When fast electrons or other charged particles are suddenly slowed down upon close encounters with other charged particles, radiation can also be emitted to satisfy the law of conservation of energy. The radiation is known as bremsstrahlung (braking-radiation) and occurs in the β-radiation of radioactive decay. It has also been detected in lightning discharges and plasmas.

1.8 THE LIMITS OF HUMAN VISION

It has already been mentioned that some forms of luminescence are faint and close to the limits of the unaided eye. It is therefore useful to pay some attention to the mechanism and perception of visible radiation.

Figure 1.17a shows a diagrammatic section through a human eye. Light passes through four principal media before contacting the retina where there are light-absorbing pigments. These media are the cornea, the anterior chamber, lens and vitreous

(a)

vitreous body

retina

cornea

anterior chamber

lens

(b)

nc   n   i   o   e

10μm

(c)

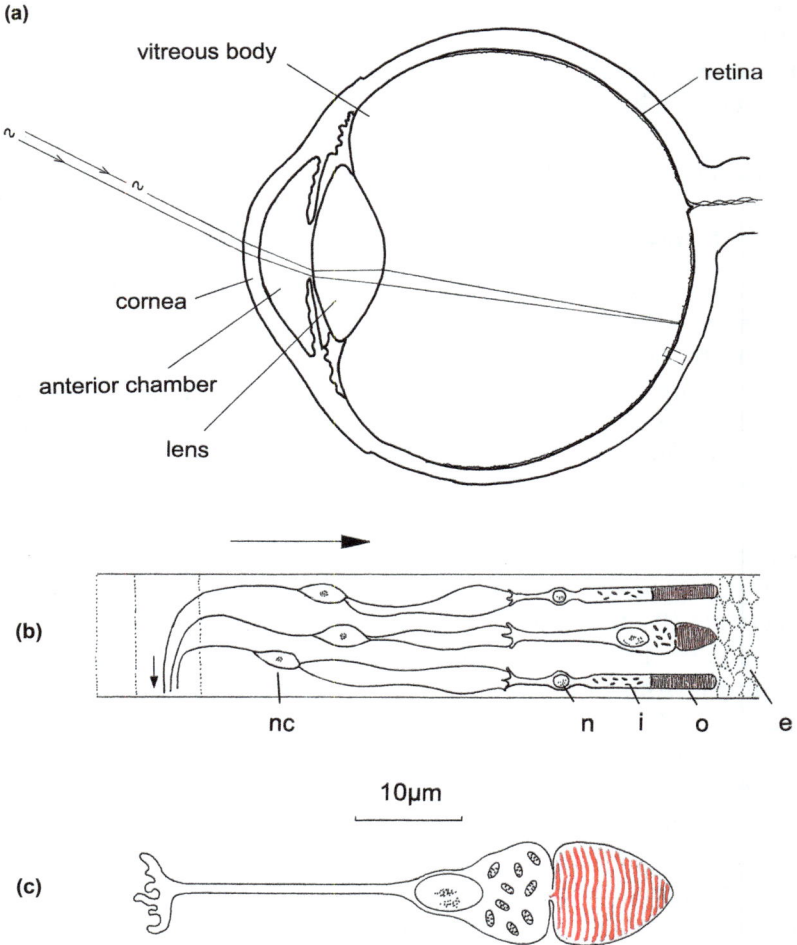

Figure 1.17 (a) Section through a human eye showing the main features. Two rays of light are shown entering the eye from a distant source. (b) A section through the retina indicated in (a) by the rectangular box lower right. Two rods and one cone cell are shown. The light path is indicated by a large arrow. At the outer segment membranes (o) photons are absorbed by the photosystem. e, epithelium; i, inner segment where chemical energy is generated to operate an ion pump; n, nucleus; ns, cells of the nervous system connecting to the optic nerve (small arrow). (c) Isolated cone cell showing detail of the light-absorbing system in the outer segment.

body. At the retina, radiation is converted into nerve impulses. There are two lenses in the eye, the cornea where most of the focussing is achieved and the lens itself where finer adjustments are made. A small part of the retina opposite to the lens, and called the fovea, is capable of the highest resolution. This region is covered in a mosaic of cone cells while most of the retina consists of a mixture of cone and rod cells.

The cones (Figure 1.17b and c) respond to bright light and there are three types that are receptive to photons of different energies. The output from the cones is responsible for colour vision. Rods have a similar form to the cones (Figure 1.17b) but are much narrower and contain a different pigment. They come into play in faint light but they do not allow colour vision. Rods are 200 times more sensitive than cones and their maximum sensitivity is at 555 nm (green light). Both structures can be seen to reside some distance back from the surface of the retina. This is to help protect, nourish and recycle them.

Light intensity is best understood in terms of the number of photons of visible light passing through an area of one square metre in one second. For bright sunlight this is a very large number, around 10^{21} photons. The human eye is capable of seeing objects over at least 14 magnitudes of light intensity (that is, a range from one to 10^{14}).

The eyes of most people respond to the light spectrum in the wavelength region 380–700 nanometres, running from violet to red. However this varies in different people and experiments have shown that some individuals can detect radiation covering the range 310–1100 nanometres, from the near ultraviolet to the near infra-red. The eye is not completely transparent to ultra-violet rays, and some of the absorbed UV photons are suspected of causing fluorescence in the eye, giving the impression of visibility in this region. Whatever the cause, it is clear that defining light in the 'visible' part of the radiation spectrum is subject to some uncertainty.

The lowest level of light that can be detected by the human eye is known as the absolute threshold. There have been many attempts to determine this intensity. Again it has been found to vary between individuals and the nature of the light. Point sources of light such as stars give different values to diffuse light such as an object viewed through frosted glass. As noted above, different areas of the retina also vary in their sensitivity to light. The region considered the most sensitive in the eye is a zone about twenty degrees away from the fovea where the rod cells are usually most abundant. The eye also needs to be dark-adapted so that the rods become fully functional. This takes 40 minutes or more in the dark.

In the Northern Hemisphere a simple test to determine the relative sensitivity of the eye is to count the number of visible stars in the open star cluster known as the Pleiades. Average dark-adapted eyesight will reveal six of them on a clear moonless night. Keen eyesight will reveal several more and the faintest of these stars will allow about 5000 photons of light to enter the eye per second. More sophisticated studies have shown that a threshold of about 10 photons can be detected by the eye from a point source during a brief flash. In this case the signal would be restricted to a small area of cones connected to a single nerve fibre. However there are complications caused by the absorption and scattering of photons within the eye prior to their interaction with the rod pigment rhodopsin in the retina.

There are additional problems associated with the detection of weak light. These may result in hallucinations or imagined moving images and are of particular relevance to phenomena such as the *ignis fatuus* (Chapter 9) and radon emission (Chapter 5). They include the presence of 'visual noise' and the Troxler effect. The former is due to random electrical fluctuations in the neural pathways and is sensed as a fine granularity when seeing in the dark. The second comes into play when staring for a long time at a bright point. After a few seconds, images around the point begin to fade away.

There is also *autokinesis* – the illusion that a dim static light in the dark when stared at can appear to move of its own accord. This is thought to be due to the lack of reference points near the light enabling movement direction to be established. It is also likely that involuntary muscle movements associated with the eye may be involved.

In bright light, 'after-images' are often observed. There are several types. Positive after-images occur where the image is still seen after it has been removed. However this only occurs for a fraction of a second and is not normally noticed. It is probably a delayed response occurring between the light receptors and the brain. Negative after-images last much longer and are seen in a colour complementary to the original image. Cones respond to the bright light and if the light persists these approach saturation and become inefficient at light capture. Cones adapted to other wavelengths step in, exhibiting the complementary colour. Rods also have a slower response time than cones. If you look at a black and white grid in dim light and close your eyes, the black areas appear almost white. The rods respond to the reduced illumination of the black areas but they appear lighter owing to their higher sensitivity.

It is also possible to sense light without it entering the retina. This can occur as physical pressure is applied to the eye such as a loud noise. It is often apparent when sneezing. Stresses are set up in the retina that activate the eye's nervous system. The lights have been termed phosphenes, not to be confused with the gas phosphine to be described in Chapter 9.

FURTHER READING

P. Atkins, *Chemistry, A Very Short Introduction*, Oxford University Press, 2015.

K. Campbell, *Chemiluminescence, Principles and Applications in Biology and Medicine*, Verlag Chemie, Weinheim, 1988.

Lumen: The Art and Science of Light, ed. K. Collins and W. K. Turner, Yale University Press, London, 2024, pp. 800–1600.

J. F. Griffiths and J. A. Barnard, *Flame and Combustion*, CRC Press, Boca Raton, Fla, 1995.

N. N. Greenwood and A. Earnshaw, *Chemistry of the Elements*, Elsevier, Amsterdam, 2nd edn, 1997.

A. Kimberly and M. Watzke, *Light: The Visible Spectrum and Beyond*, Black Dog & Leventhal, New York, 2015.

J. C. Kotz, P. M. Treichel and P. A. Harman, *Chemistry and Chemical Reactivity*, Thomson Learning, Chicago, Ill, 11th edn, 2023.

Oxford Dictionary of Chemistry, ed. J. Law and R. Rennie, Oxford University Press, 2020.

M. A. Liberman, *Introduction of the Physics and Chemistry of Combustion: Explosion, Flame, Detonation*, Springer, Berlin, 2011.

J. N. Lythgoe, *The Ecology of Vision*, Clarendon Press, Oxford, 1979.

D. Macaulay and S. Keenan, *Eye: How it Works*, David Macaulay Studio, Harvard, 2015.

A. Piel, *An Introduction to Laboratory, Space and Fusion Plasmas*, Springer, Berlin, 2010.

Chemiluminescence and Bioluminescence. Past, Present and Future, ed. A. Roda, Royal Society of Chemistry, Cambridge, 2011.

J. M. Russell, *Elementary: The Periodic Table Explained*, Michael O'Mara Books, London, 2021.

E. F. Schubert, *Light-Emitting Diodes*, Cambridge University Press, 2nd edn, 2006.

G. J. Tallents, *An Introduction to the Atomic and Radiation Physics of Plasmas*, Cambridge University Press, 2018, DOI: 10.1017/9781108303538.

A. J. Walton, Triboluminescence. Adv. Phys., 1977, **26**, 887–948.

M. Weller, T. Overton, J. Rourke and F. Armstrong, *Inorganic Chemistry*, Oxford University Press, 2018.

Y. J. Xie, & Z. Li, Triboluminescence: Recalling interest and new aspects, *Chem*, 2018, **4**: 943–971.

REFERENCES

1. D. A. Frank-Kamenetskii, *Diffusion and heat transfer in chemical kinetics*, Plenum, New York, 1969.
2. A. G. Gaydon and H. G. Wolfhard, *Flames, their structure, radiation and temperature*, Chapman and Hall, London, 1960.
3. M. Faraday, *A Course of Six Lectures on the Chemical History of the Candle*, Griffin, Bohn & Co, London, 1861.
4. H. H. Wollaston, A Method of examining Refractive and Dispersive Powers by Prismatic Reflection, *Phil. Trans. R. Soc.*, 1802, **92**, 365–380.
5. D. Brewster, On a new phenomenon of colour in certain specimens of fluorspar, *Rep. Brit. Ass. Adv. Sci.*, 1838, 8th report, pp. 10–11.
6. G. Stokes, On the change of refrangibility of light, *Phil. Trans. R. Soc.*, 1852, **142**, 463–562.
7. H. G. Emeleus, Spectroscopic study of combustion of phosphorus trioxide and hydrogen phosphides, *J. Chem. Soc.*, 1925, **127**, 1362–1368.
8. E. Samuel Reich, Evidence for bubble fusion called into question, *Nature*, 2006, **440**(7081), 132.

CHAPTER 2

Spontaneous Combustion in Human Hands

2.1 INTRODUCTION

Spontaneous combustion, also known as self-ignition has been defined as the ignition of a substance in air without any external input of heat.[1] Nevertheless, the initiation of spontaneous combustion cannot be defined by a single temperature. Values quoted for auto-ignition temperatures for example are far from constant and depend upon such factors as air humidity, concentration of fuel and the time required for initiation. As a general rule however, most self-ignition phenomena have been described as occurring below about 250 °C. In this chapter I have imposed a more stringent upper temperature limit so that most of the described phenomena self-ignite at, or close to the ambient temperature (20 °C). A number of substances combust at temperatures between about 250 °C and 400 °C. These auto-ignition temperatures are still low when compared with some common flammables such as methane and hydrogen and are sometimes described as *easily combustible.* Some reference to these will also be made. Combustion is usually accompanied by visible flame and virtually all of the examples described in this book are of this type. The term has however occasionally been

Luminous Phenomena: A Story of Spontaneous Combustion, Phosphorescence and Other Cold Lights
By Allan Pentecost
© Allan Pentecost 2025
Published by the Royal Society of Chemistry, www.rsc.org

extended to cover a wider range of oxidation processes and a few examples of these are also given.

There is a long record of spontaneous combustion events occurring in manufactured substances, ranging from some of the chemical elements and simple compounds to more complex but less well known materials found in the laboratory. In their efforts to synthesise new compounds, chemists have reported large numbers of such substances. Most were discovered by accident, such as the organometallic compounds. In this chapter a range of examples drawn from different elements and compounds have been chosen to illustrate their diverse constitution and chemical action. The phosphorus hydrides will however be reserved for consideration in Chapter 9. Many of the substances described in this chapter are unstable in air and rapidly react with the oxygen within it. We therefore begin by first taking a look at chemical reactivity and its measurement.

2.2   CHEMICAL REACTIVITY AND ELECTRONEGATIVITY

As chemists began to understand how the chemical elements behaved it became clear that there existed patterns of reactivity. It was found for example that the alkali metals such as sodium and potassium were very reactive with non-metals such as oxygen, while metals such as gold, were almost non-reactive towards most other elements. Eventually measures of reactivity were devised, allowing the elements to be compared. One of these measures, called electronegativity, although imprecise, effectively demonstrated the different reactivities of the elements and many of their compounds.

Electronegativity is essentially a measure of the willingness of an atom to accept an electron. Strongly electronegative atoms have a small radius and an electronic structure just short of a stable configuration. An example is the element chlorine. Chlorine is short of a single electron to complete a stable shell. This makes it extremely reactive and capable of extracting an electron from many of the less electronegative elements. During the process, heat is frequently produced. In contrast, sodium atoms are more than fifty percent larger than chlorine atoms and contain only one electron in their outer shell. It is difficult to make a sodium atom accept another electron, but the single outer electron of sodium is vulnerable to capture and is easily lost. The electronegativity of sodium is so low that it and similar elements are usually termed electropositive. When an atom of chlorine contacts an atom of sodium, the single outer electron of sodium is easily removed. A violent reaction ensues and much heat is produced. The result is common salt, sodium chloride.

You may remember from the previous chapter that atoms are electrically neutral – they contain the same number of positively charged protons and negatively charged electrons so the electric charges cancel out. We can see that once a chlorine atom has removed an electron from a sodium atom, this atom is no longer electrically neutral. Losing a single negative charge leaves the sodium atom with a positive charge. In the same way, the chlorine atom gains an electron and becomes negatively charged. The attraction between the positive and negative charges remains, but the charge now rests upon two atoms which are effectively bonded together. The charged atoms are called ions, and the bond between them is called an ionic bond. Ionic bonds tend to be formed by strongly electronegative atoms and weakly electronegative (ie electropositive) atoms. However there are many chemical reactions involving atoms of intermediate or equal electronegativities. In most of these cases, the electron is attracted to the other atom but still spends some of its time with the parent atom. Bonding of this type is common and known as covalent bonding.

Combustion involves reactions between strongly electronegative elements such as oxygen and elements or compounds containing more electropositive atoms. Examples of the latter are hydrogen and carbon. The process normally requires an initial source of energy to start the reaction and this is called the activation energy. During chemical reactions, bonds are strained and then broken and this requires an input of energy. Once the reaction has been completed and the new products are formed, energy is no longer required. After a reaction has started it can be self-sustaining with energy being released as heat as the new bonds are formed.

If we confine our attention to substances that can be observed under controlled conditions, we find that the number of spontaneously combusting substances is small when compared with the vast number of chemical compounds that are currently known. With a knowledge of the reactivity of the chemical elements, we should expect the self-igniting elements and compounds to be listed among those of weak electronegativity, namely the strongly electropositive atoms of the Group 1 alkali metals. All of these elements have a single, vulnerable electron in their outer shell. As the Group 1 column of the periodic table is descended, reactivity increases. One might expect the reverse to be the case. For the highly reactive element caesium, near the bottom of Group 1 the large nucleus of caesium must surely produce a strong pull on the oppositely charged electrons, holding them in their orbits. The reverse occurs for two reasons. First, the outer shell electron is on average a long way from the nucleus. The attracting strength of the nuclear charge falls off rapidly with distance so this electron does not experience a strong inwards pull. Second, the other 54 electrons of caesium provide a 'mist' of negative charge between the nucleus and the outermost electron, shielding it from the charge on the nucleus. This makes the outer electron of caesium vulnerable to capture by another atom. A useful measure of the ability of an electron to remain with its atom is its ionisation energy. This is the energy required to extract an electron from the atom to a sufficient distance that it is no longer influenced by it. As the Group 1 elements of the Periodic Table are ascended, the outermost electron gets closer to the nucleus and the number of shielding electrons gets smaller. It should not be surprising to find that the ionisation energy of the outer electron of sodium is 5.14 eV and that of caesium is considerably less at 3.89 eV making the latter the more reactive element.

In Figure 2.1, the electronegativity of the elements is plotted in increasing order with the least negative, that is, electropositive elements, on the left and the most electronegative on the right. It is clear that most of the strongly electropositive elements spontaneously combust but the others are scattered throughout the table. With the exception of Groups 1 and 2, there is no clear correlation between self-ignition in air and electronegativity. Evidently other factors must come in to play. Among them, the physical state of the element is important. For elements that are solid at room temperature their state of division must be considered. When in a finely divided form, a large surface area is exposed to the air allowing greater access to oxygen. Some

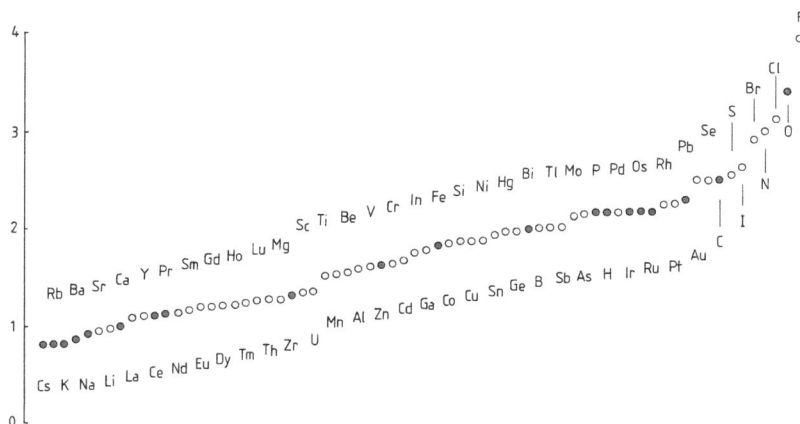

Figure 2.1 Electronegativities of the elements in ascending order with the least electronegative (= electropositive) elements at the left. Elements that are known to spontaneously combust in air are shown in red. Elements shown in yellow are not known to be spontaneously combustible but some of their simple compounds are. Oxygen coloured blue.

elements are easily reducible to a fine state of division whereas others are not. Extremely fine division into 'nanoparticles' is now possible for many materials.[2] These particles are typically a few nanometres (10^{-9} m) across and present a huge surface area, each particle containing just a few hundred or thousand atoms. Exposing the solid elements in this state to air or oxygen would probably modify considerably our current knowledge of their combustion properties – and some of their compounds.

Another factor affecting self-ignition is the degree to which films of oxide or nitride build up on the surface during contact with air. The physical form of these films differs from element to element. Aluminium for example is a reactive metal but becomes covered with a strong film of oxide preventing further oxidation. Water vapour, invariably present in the air, also plays a part in the formation of these films. By way of contrast, in elements such as sulphur the oxides are volatile and are unlikely to significantly interfere with their oxidation.

We shall now turn our attention to the element oxygen, the component of the air that causes spontaneous combustion. This will be followed by an examination of spontaneous combustion of some of the elements themselves before progressing to compounds and mixtures. In each case examples will be chosen to illustrate the range of processes occurring and their interrelationships.

2.3   OXYGEN – DISCOVERY AND ORIGIN

Interest in the atmospheric gases extends back several hundred years but it was not until the 1600s when the Polish alchemist Michael Sendivogius heated potassium nitrate, a component of gunpowder, to yield this gas as the salt decomposed. In 1754 the biologist Charles Bonnet noticed that bubbles of gas developed on the illuminated leaves of the grape vine when they were immersed in water. Twenty-five years later, Jan Ingenhausz, after a visit to the chemist Joseph Priestley found that the bubbles were of oxygen. Soon after, the Swiss biologist Jean Senebier[3] employed some ingenious experiments and found that the amount of oxygen produced by leaves was approximately the same as the amount of carbon dioxide they absorbed. He also found that the process occurred in the green parts of the leaves and thus discovered photosynthesis.

Up to 1780 oxygen was not regarded as an element due to the widespread acceptance of the 'Phlogiston theory'. The gas was then known as 'dephlogisticated air'. This concept led to much confusion but did not deter Henry Cavendish from undertaking some detailed experiments on the composition of the atmosphere.[4] Cavendish, like several other chemists at the time, was aided by a device called an eudiometer. This apparatus allowed the actual amounts of different gases taking part in chemical reactions to be measured. Its basic components were an inverted glass tube filled with water or mercury with the open end immersed under a dish of the same liquid. The name was borrowed in part from the Greek language and means 'goodness of air meter' (see Box 2.1).

During the 19th and 20th centuries, methods of measuring oxygen in the atmosphere became increasingly sophisticated allowing different regions of the atmosphere to be analysed and compared. By the 21st century methods gave a precision of just a few parts per million using instruments such as fuel cells and mass spectrometers.

BOX 2.1   EXPERIMENTS ON AIR UNDERTAKEN BY HENRY CAVENDISH

Henry Cavendish was born into a wealthy aristocratic family in 1731. His father took an interest in science and brought him to meetings of the Royal Society in London. Henry was attracted to the scientific instruments on display and soon became established in London's scientific community. This led, among other things, to a study of gases, and by the 1780s he was undertaking experiments on air using his own eudiometer (Figure 2.2). This consisted of a bent soda-glass tube (M) which was carefully filled with mercury and then immersed under more mercury held in the cups as shown in the figure. A known volume of air or other gas was then added to the bent tube using the capillary tube ABC. It was filled with the desired amount of gas and then up-ended with a finger placed over the end C so that the gas could enter the tube M. Following the gas, a liquid reagent could be added to M in the

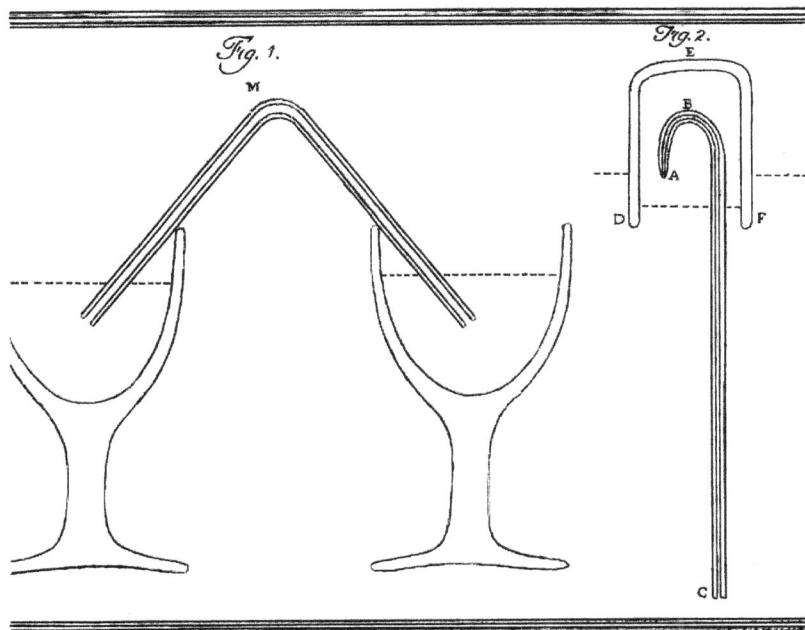

Figure 2.2 Original diagram of the eudiometer used by Henry Cavendish in his studies on the composition of the air. See text for details.

same manner where it would contact the gas in the tube. Chemical reactions between the liquid and gas could then take place and any volume change measured. Reactions within the gas could also be initiated using an electric spark. This was achieved using an electrically charged sphere placed close to the tube M on the outside. A discharge from the sphere to the mercury within the tube raised the gas to a high temperature allowing reactions to occur between gas mixtures. By this means Cavendish managed to show that the amount of 'dephlogisticated air' was close to 20% by volume. It was left to the French chemist Antoine Lavoisier to clarify the situation and coin the term 'oxygen' for dephlogisticated air.

Once oxygen was discovered, the question of its origin in the atmosphere was soon raised. It now appears that almost all of it is the result of photosynthesis. It was initially assumed that its concentration gradually increased over time to its present level but research by geochemists suggests that oxygen levels have varied considerably in the past.[5–8] The situation is complicated by the presence of other gases such as carbon dioxide and the reaction of oxygen with components of the earth's crust, particularly iron sulphides. These sulphides are common minerals and are readily converted into iron oxides in the presence of oxygen and water. This leads to the removal of oxygen from the atmosphere. Taking into consideration the distribution of land masses and other contributing factors, it appears that the oxygen concentration reached a maximum of about 30% in the late Palaeozoic around 270 million years ago.[5] It then declined before increasing again more or less regularly from the late Triassic to the present day. One model suggests that the oxygen concentration fell to around 10% in the early Jurassic. Unfortunately these estimates do not always tally with the fossil evidence. A study of fossil charcoal resulting from ancient wildfires suggested that during the 'low oxygen' periods predicted by some models, fires were in fact taking place. These fires would have been caused mainly by lightning strikes. A series of experiments showed that sustained combustion of wood is not possible in an atmosphere containing less than 15% oxygen,[6] at variance with some of these models.

It is easy to forget that the atmospheric concentration of oxygen, along with the other gases, is dependent upon the pressure and thus the height above sea level. At or near sea level, the present concentration is close to 21%, but in relative terms it is much lower in the high mountains due to the fall in atmospheric pressure. At Gorak Shep, just below Everest Base Camp at an altitude of 5160 m, wood fuel is difficult to burn. Instead it tends to smoulder although butane gas burners still operate. Here, the oxygen concentration is about half that at sea level. Even at 3600 m altitude, burning wood can be an issue.

These observations are clearly important when considering spontaneous combustion. Today the earth's atmosphere at or near sea level varies little in its oxygen content despite processes that continuously produce and absorb it, such as daytime photosynthesis (production) and night-time respiration (removal). A substance that self-ignites at sea level may not do so at the top of a mountain.

While the atmospheric gases as a whole have a fairly predictable concentration, varying mainly with altitude, local variations in soil and water can be high. Oxygen concentrations in the soil, especially when saturated with water can easily approach zero owing to the respiration of plant roots and microbes. Conversely, when rates of photosynthesis by water plants are high, the concentration of oxygen can rise dramatically. Bubbles of oxygen can often be seen on the submerged leaves of plants where they rise to burst on the water surface. Bubbles caught within algal mats can remain for extended periods providing locally high concentrations of oxygen that could approach 100% (Figure 2.3). Nutrient-rich lakes that contain an abundance of planktonic

Figure 2.3   Bubbles of oxygen bursting through an algal mat on a warm summer's day. Image 30 mm wide.

algae are often over-saturated with oxygen leading to bubble formation. In shallow waters blown by a wind, lines of bubbles and floating debris often collect in 'windrows' due to circulating water currents. Locally high levels of oxygen are again to be expected above the rows although they would soon be dispersed by the wind. Under forest canopies photosynthesis and respiration cause the oxygen concentration to vary by up to about 50 parts per million from day to night.[9] Nonetheless this is a minute change compared with its actual concentration.

Auto-ignition temperatures are known to vary with the oxygen concentration in the atmosphere.[10] For the materials so far tested, this temperature usually falls as the oxygen content rises. Nevertheless materials do vary in their response. Some hydrocarbons for example have an ignition temperature in pure oxygen that is similar to that in air while in others a far lower temperature is observed. Most of these experiments have been conducted on pure compounds and it is unclear how higher concentrations of oxygen would affect many natural materials.

2.4   SPONTANEOUS COMBUSTION OF THE METALLIC ELEMENTS

In this section we look at those metallic elements known to self-ignite. We begin with the Group 1 elements. This group contains the least electronegative elements, all of which react vigorously with oxygen. This group also includes hydrogen, a gas whose chemical properties are at variance with those of the other Group 1 elements all of which are metals. Hydrogen will therefore be discussed in the section dealing with the non-metallic elements. After examining Group 1 we move on to the Group 2 elements, working our way through the Periodic Table from left to right.

2.4.1   Alkali and Alkaline Earth Metals

Of the naturally occurring elements, 66 can be regarded as metallic. Of these a considerable number have been reported to be spontaneously combustible in air. We begin by looking at the alkali metals, a group of six elements. They belong to Group 1 at the left hand end of the Periodic Table. In order of atomic number they are lithium, sodium, potassium, rubidium, caesium and francium. All are extremely reactive towards oxygen and water but only the heavier members combust spontaneously. Some of the lighter members however, such as sodium and potassium catch fire in the presence of water. Sodium dropped into water reacts violently to produce sodium hydroxide and hydrogen. The resulting high temperature causes the metal to melt, and being light, it floats. When the metal has been reduced to a small globule it often ignites. Sodium is cast into sticks and kept under oil to prevent attack by both oxygen and water. When the sticks are removed from oil a thin film of hydroxide rapidly forms over the surface, slowing down the oxidation. In this case the heat produced by the reaction is carried away quickly by the large volume of metal in the stick.

Potassium is more reactive than sodium and is difficult to work with. Like sodium, the metal must be stored under oil to prevent oxidation. In 1806, Humphrey Davy placed a small piece of moistened potash (potassium hydroxide) on electrically insulated platinum foil. He connected the foil to one battery terminal and a platinum wire to the other. Applying the wire to the potash, he found .... *'a vivid action, the potash began to fuse and there was violent effervescence at the upper surface...and small globules with a high metallic lustre appeared ... some of which burnt with explosion and bright flame...'*. Davy was said to have jumped with joy at his discovery – he discovered sodium a few days later. If a small piece of potassium about the size of a pea is dropped into water, a violent reaction ensues. The potassium begins to burn almost immediately with a characteristic bright purplish flame, often accompanied by a small explosion. The explosions are caused not by the metal itself, but by the production of hydrogen. This mixes with the oxygen of the air and is heated to its ignition temperature by the burning and now molten potassium. The metal is raised to a high temperature by its reaction with water and then burns in the oxygen of the air. The heat cannot escape rapidly enough to prevent combustion. Although water is a good conductor of heat, potassium floats on water and is probably insulated from the water beneath by a cushion of hydrogen gas. Potassium can also be made to burn in the absence of water. If a small piece of the metal is gently squeezed, it liquefies and soon catches fire in the air. Interestingly, this experiment fails in completely dry air, so water vapour must play a part in the combustion reactions.

The remaining members of the Group 1 metals, rubidium, caesium and francium are much more reactive than potassium. Rubidium and caesium are spontaneously flammable in air. They burn in air as small slivers or wires and they do not require compression. Some years ago I had a caesium lamp constructed for a research project. The lamp builder had to make three attempts to insert the metal in the lamp before it caught fire in the air. It is one of the most difficult of all the stable, that is, non-radioactive elements to handle.

At the very base of Group 1 sits the element francium. This should be even more reactive with oxygen than caesium but it is highly radioactive. It has a short lifetime and little is known about its chemistry. It would almost certainly catch fire in air if sufficient material could be isolated. There is another group 1 element, lithium that sits above sodium. It is the least reactive of the group and does not share many of the properties of the other alkali metals. Lithium has a small atom and its ionisation energy is higher than that of sodium. With its small number of shielding electrons, lithium holds onto the outer valency electron more strongly than other members of the group.

The alkaline earth metals of Group 2 provide another set of closely related metallic elements. They are also highly reactive towards oxygen. Like the alkali metals, they cannot remain stable in the presence of air and two of them, calcium and barium are known to combust when powdered, the latter with a brilliant green flame. Strontium is also likely to do so. Radium, the least electronegative element of the group is also reactive, but is difficult to handle as it is radioactive.

2.4.2   The Lanthanoids

The lanthanoids are a large group of metallic elements that share such similar properties that they could not be separated from one another in significant amounts until the middle of the 20th century. Also known as the 'rare earths' they are thinly spread through the earth's crust with few minerals sufficiently enriched for their profitable recovery. The lanthanoids occupy a special position in the Periodic Table owing to their unusual electron configuration. They possess low electronegativities in common with the Group 3 elements headed by the rare metal scandium. They should therefore be expected to be easily combustible. Among them, lanthanum, cerium, and perhaps praseodymium appear to be the only elements that can spontaneously combust as fine dusts, although all of the lanthanoids are reactive towards oxygen. Interestingly the most reactive member, europium, has a higher autoignition temperature than some of the others in the group such as cerium. However as a dust, europium is considered an explosion risk in air.

Several lanthanoids burn at a temperature of around 150 °C and before most of these elements were isolated they were being put to use as mischmetal (mixed metal) in cigarette lighters. Cerium is the main component and is still used to make sparks for lighting fires. It was discovered and named by Carl Auer von Welsbach who first obtained it from the mineral monazite, a rare phosphate of mixed lanthanoids. In the 1890s monazite sand was being used as a source of thorium to make gas mantles but Welsbach found that the residue could be reduced to an alloy with pyrophoric (fire-producing) properties. By scratching the surface with a harder metal, good sparks could be made. Upon abrasion, fine particles of the mischmetal were released and ignited as a result of the frictional heat generated during contact with a harder surface. Welsbach subsequently founded the Treibach Chemical Works in Austria in 1908 for the large scale production of the alloy. Iron was added later to make the product more durable and mischmetal, also known as ferrocerium, has been used to generate sparks for lighting fires ever since. The Ronson Company began making cigarette lighters in 1910 with a mischmetal sparker and naphtha as fuel. The company was founded by Louis Aronson who was also an inventor of safety matches. In 1926 he produced the 'automatic operation Banjo Lighter' which could be ignited and extinguished by two simple hand movements. It was an immediate success (Figure 2.4). Mischmetal consists mostly of cerium (c. 50%) with about 20% lanthanum, 5% praseodymium and smaller amounts of other lanthanoids, strengthened with iron oxides. Spontaneous combustion does not occur in these lighters because frictional heat is required to initiate the oxidation. Many lighters now use piezoelectric sparkers consisting of a ceramic lead compound. Thorium, a radioactive element in the related actinoid series, has also been described as close to pyrophoric.

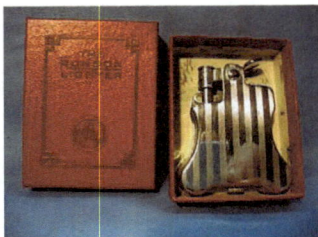

Figure 2.4   The Ronson Banjo Lighter and case. The first successful lighter equipped with a mischmetal flint for ease of operation. Image by Peter Aythorpe.

Zirconium and hafnium belong to Group 4. They have special applications in alloys. In a finely divided state, with a grain size of about ten micrometres both have been known to catch fire. Special photoflashes have been devised using a combusting mixture of zirconium powder and oxygen difluoride, OF_2.

2.4.3   Transition Elements

These elements sit in the middle of the periodic table. They are all metals and are characterised by their several oxidation states, their ability to form complexes with other elements and their frequently brightly coloured salts. Those exhibiting self-ignition behaviour include iron, ruthenium, osmium, iridium, copper and zinc.

Iron is an element which most of us recognise as chemically reactive. Iron objects exposed to the air soon rust as a result of a chemical reaction between the metal and the oxygen in the atmosphere. The process is more complex than was first thought and water plays an important part. Metallic iron is nevertheless less reactive toward oxygen than the Group 1 and Group 2 elements. However, if a large area of the metal is exposed to the air, it is soon oxidised. Steel wool for example when moistened and mildly acidified can remove oxygen in air in about 20 minutes within a closed container.

It has been reported that iron reduced electrically often steams when rinsed owing to its rapid oxidation. In one case a large detached fragment was seen to glow and proceed to burn through the French polish on a bench. As a finely divided black powder, iron has been seen to incandesce when dried due to its oxidation. There have been attempts to control the combustion of iron in air by preparing it as nanoparticles. It might then be used in this form to fuel internal combustion engines with the aim of reducing the release of greenhouse gases. It is now possible to prepare nanoparticles of iron and other metals in a range of sizes to give optimal performance as a fuel. Using iron in preference to petrol has an advantage since on a volume basis, iron releases nearly twice as much energy. Unfortunately it is much heavier than petrol and as a fine powder could be hazardous to handle. Upon oxidation, hydrogen, the preferred alternative, provides twelve times as much energy as iron on a weight by weight basis. There is a rapidly expanding market for nanoparticulate materials and they have found use in a wide range of industries from antimicrobials, delivery of drugs and as catalysts in biotechnology.[2] They are normally used as suspensions and are thus immune from combustion.

Other elements in the iron group have also been shown to ignite. These include powdered ruthenium and osmium. Along with iridium these metals are widely regarded as unreactive and their ability to combust may appear surprising. It is probably connected to their physical state when finely divided. Finely divided copper produced by precipitation in solution has also been reported to combust when dried.

Zinc is a familiar metal and is easily vaporised at 400 °C. It then condenses as fine dust. The dust is highly reactive and often employed in chemical syntheses as a cheap but powerful reducing agent. However it has not been reported to self-ignite. This is probably because the particles are soon covered in a thin film of oxide. In fact zinc dust is used as a component of solid rocket fuel where spontaneous combustion would be a serious issue.

2.4.4 The Group 14 Metals

The Group 14 metallic elements are germanium, lead and tin and are to be found towards the right hand end of the Periodic Table. A significant feature of the group is the presence of four electrons in the outer shell of the atoms. These are able to enter into bonding with other elements and in the case of tin and lead, compounds occur where either all four electrons are involved or just two.

Finely divided tin can be made by placing a piece of zinc metal in a solution of a tin salt, such as tin chloride. Once dried, it catches fire on mild heating. Finely divided lead is more difficult to obtain in this way but it has been achieved by heating lead stearate in an alcohol called octanol. The lead forms nano-particles but they are coated in an organic film preventing combustion. Free of this film there is good reason to believe that lead can self-ignite as will be seen later. The remaining Group 14 elements are non-metals and are described below.

2.5   THE NON-METALLIC ELEMENTS

2.5.1   Hydrogen

This element is usually found in Group 1 of the Periodic Table along with the alkali metals. Because hydrogen has a single electron whose orbit is close to the nucleus its ionisation energy is much higher than that of the other Group 1 elements giving it a higher electronegativity. It is placed in Group 1 on account of its electronic configuration.

Hydrogen is the simplest of all elements and the most abundant in the universe. Best known as a constituent of water, this gas is not present in the atmosphere in significant amounts. Owing to its lightness, it has been employed for many years in balloons and dirigibles. The auto-ignition temperature of hydrogen is about 500 °C so it would not be expected to catch fire at room temperature. However, there are several reports that seem to contradict this. Two of these are presented below.

2.5.1.1   Craigmillar. In March 1982, a street lighting inspector noted a small fire at Craigmillar Quarry in Scotland. Driving closer he saw a white flash the height of a three storey building, from the top of which spurted rocket-like flames in many directions. Then he felt a tremor caused by an underground explosion. Later in the day a large crater was discovered in the quarry, which was being used as a tip. Traces of the explosion were recovered up to 2 km away. Several years previously the quarry had housed a firework factory and contained a residue of firework chemicals. A decision was made to bury it in a concrete chamber within the quarry floor. The chemicals were placed in steel drums and later the chamber was covered with rubbish and finally earthed over. An inspection took place subsequent to the explosion and three suggestions as to its origin were put forward: (1) gas produced by the chemicals could have escaped and been ignited by the many small fires that occasionally burnt on the tip; (2) spontaneous ignition occurred within the dump by the reaction between firework chemicals, and (3) hydrogen had been formed through the action of acidic water on aluminium and magnesium powders – often used in fireworks. The gas had then escaped the chamber under pressure, mixed with methane produced in the dump and ignited. A public analyst thought that the third possibility was most likely but it was unclear how acidic water could have entered the chamber. To begin with, the alkaline nature of concrete would surely have removed any acidity. However aluminium powder is notoriously reactive towards strong alkalis and this may be the source of the hydrogen. Also it is not clear how fire could have spread into the chamber if it started outside.

2.5.1.2 The Hindenburg. The Hindenburg disaster of 1937 received enormous publicity mainly as a result of the event being filmed in its later stages. This huge airship was held aloft by a chambered hydrogen balloon but caught fire moments before it docked to its mooring mast in New Jersey in the United States. The cause of the tragedy remains a mystery although the most likely explanation is the ignition of hydrogen by an electric spark. The accident happened a few minutes after a mooring rope contacted the ground and it is possible that this resulted in a flow of electricity between the airship and the earth. This could have been facilitated by rain making the rope wet and electrically conducting. There were eye-witness statements indicating that part of the airship's cover was flapping suggesting the escape of gas, possibly the result of a leak caused by some structural damage during manoeuvering. While this remains the most plausible explanation, several other hypotheses have been put forward, but the possibility of spontaneous combustion of the hydrogen appears to have gone unreported. Could the leaking gas have self-ignited? Seven years earlier the British airship R101 also crashed and caught fire with great loss of life. Again the cause of the fire is unknown but the spontaneous ignition of calcium flares on contact with rainwater, which were in use at the time, is one possibility. Of further significance are the fires that have destroyed zeppelins while they were being filled with the gas.

2.5.1.3 Catalytic Action. The increasing use of hydrogen as a fuel has led to experiments to determine whether the gas is really spontaneously combustible. They appear to show that pure dry hydrogen forced through small holes will not spontaneously ignite. However if the gas contains iron oxides in the form of dust it may do so. Several hypotheses ensued. For example, the dust surface could have reduced the iron oxides to the form of pyrophoric iron. In contact with air, this material would ignite. Further experiments could not confirm this, but indicated that instead the phenomenon was electrical. As the suspended dust contacted the sides of the container, friction caused the particles to become electrically charged. Then as the gas exited the holes, an electrical discharge occurred igniting the air-hydrogen mixture. Hydrogen is unusual in many respects. As a gas, it shows the Joule–Thompson effect. This leads to a change in the temperature of a gas when it is released through a small hole. With most gases, the temperature drops but with hydrogen it increases, but at room temperature the change is much too small to permit ignition.

Another feature of hydrogen is its ready oxidation in the presence of a catalyst. Catalysts accelerate chemical reactions, without directly taking part in them, although they undergo reversible changes during the process. One of the first catalysts to be discovered was platinum black. The remarkable properties of this material were described in the early 1800s by Edmund Davy. He found that *'to produce heat it is only necessary to moisten any porous substance, such as sponge. with alcohol or whis-key, and to let a particle of the (platinum black) powder to fall on the substance so moistened. It immediately becomes red hot and remains so until the spirit is consumed'.* He also discovered that a mixture of oxygen and coal gas could react together in its presence. Soon after, Johann Doebereiner found that a mixture of hydrogen and air passed over spongy platinum, a form of platinum black, ignited spontaneously. With the help of E. Furstenberg he produced a self-igniting lamp (Figure 2.5). The hydrogen was produced by the reaction of zinc and dilute sulphuric acid, the latter being contained in an outer glass cylinder. Gas production was controlled by means of a valve which stopped the reaction when the pressure inside the lamp reached a particular value. The lamp proved popular through most of the 19th century but Doeberiner made little money out of it as it was not patented.

Platinum black is finely divided platinum metal with a large surface area. Oxygen molecules are attracted to the surface and the atoms become attached. Here, their chemical activity towards hydrogen is greatly enhanced. When hydrogen gas is played upon the surface its rate of oxidation to water becomes so great that the sponge ignites the gas. Platinum wire can produce similar results with other gases. An early chemical entertainment consisted of a spirit lamp filled with alcohol and a wick supporting a coiled platinum wire. The lamp was lit allowing the wire to glow. Then the flame was extinguished. The wire continued to glow as the alcohol was oxidised to acetaldehyde until all was consumed. Sometimes Eau de Cologne was added to fill a room with scent. There may be other, so far undiscovered, catalysts capable of igniting flammable gases such as hydrogen in the presence of air. Under the right conditions, hydrogen can therefore self-ignite.

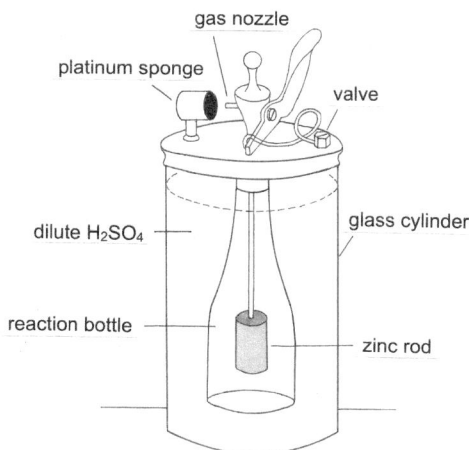

Figure 2.5 The Doebereiner Lamp.

Hydrogen, like many other elements, can combust in gases other than oxygen. One of the most dramatic is the reaction between hydrogen and chlorine. The two gases may be mixed safely in the dark but strong light causes a violent explosion. The first recorded explosion of such a mixture occurred in 1809 when the French chemists Gay Lussac and Louis Thénard were experimenting with them, noting that the reaction occurred slowly under some conditions.[11] This was easy to observe as the pale green colour of the chlorine gradually disappeared from the flasks as the reaction took place. Realising that light played a part, some containers were illuminated by direct sunlight when *they suddenly inflamed with a loud detonation, and the jars were reduced to splinters.* Much research followed and by the early 1900s it was concluded that light must be responsible for breaking the chlorine molecule (Cl_2) into highly reactive chlorine atoms. These attack hydrogen molecules (H_2) to form hydrogen chloride (HCl) and a reactive hydrogen atom. The reactive hydrogen then splits another chlorine molecule liberating more chlorine atoms and so on. A rapid chain reaction ensues and an explosion results.

It is now known that the highly reactive chlorine atoms are free radicals – atoms containing a single unpaired electron in their valency shell. Interestingly the reaction does not take place if the two gases are thoroughly dried and it is known that water vapour acts as a catalyst in the reaction. It is also known that only blue light can initiate the reaction. This has been nicely confirmed by thermodynamic measurements. The bond energy of a chlorine molecule is 242 kilojoules per mole, close to the energy of blue light of wavelength 494 nanometres allowing the bond to be broken. Photons of red, green and yellow light have insufficient energy to break the chlorine bond so the mixture does not explode.

2.5.2 Carbon

There have been occasional reports of carbon as wood charcoal spontaneously combusting. Charcoal is a fairly pure form of carbon and it is well known for its high surface area to volume ratio thus exposing a large number of its atoms to the air. Carbon is reactive towards oxygen and this combined with its low heat conductivity would make combustion favourable. The combustion of carbon will be dealt with more fully in the next chapter.

2.5.3   Phosphorus, Bismuth and Antimony

Two elements in Group 15 can self-ignite, phosphorus and bismuth. A third, antimony is also notable. Bismuth and antimony have some metallic properties but phosphorus is a non-metal.

Phosphorus is considered by many chemists as the most remarkable of all elements. This 'bringer of light' was discovered by the alchemists several centuries ago. The early explorers of chemistry could only produce small amounts at a time and it remained both expensive and obscure for a long period. As one of the chemical elements, phosphorus consists of one kind of atom, but it is an atom with some interesting properties setting it apart from all others. One property, previously noted, is that it can glow in the dark – a feature that was to make large sums of money for entrepreneurs in the years after its discovery. The mechanism responsible remained elusive until 1974. It is caused by the presence of short-lived HPO and P_2O_2 molecules that undergo excitation *via* oxidation with consequent light emission.

Accidents with white phosphorus were reported as early as the 17th century. Small fragments found their way onto clothing and into people's pockets causing fires. The reactivity of this form of the element is attributed to its structure. The atoms are bonded into four-cornered figures that look like pyramids. To maintain their shape, the molecules are strained and this is the reason why white phosphorus is so reactive. Red phosphorus is a form that does not have this kind of structure. It is much more stable and is safer to handle.

It was soon appreciated that the inflammable nature of phosphorus could have useful applications in lighting fires. In Italy the chemist Peyla enclosed a taper with a frayed end into a sealed glass tube with a small piece of white phosphorus at one end. If this end was dipped into hot water and the tube broken open, the taper, which had absorbed some of the melted phosphorus, would catch fire as it gained access to the air. A little later, in the 1780s, a variant called the 'pocket luminary' was invented. This consisted of a small bottle whose inside was coated with phosphorus. To use, the bottle was opened and a sulphur match was inserted to scrape a little phosphorus off the inside of the bottle. This was then rubbed against the bottle's cork generating sufficient heat to ignite the match.

The 18th century pocket luminary was soon overtaken by more convenient devices that became known as 'lucifers'. Here the match was ignited by rubbing the head against a rough surface. These matches were no less a fire hazard and some formulae still contained white phosphorus. Their manufacture was carried out on a huge scale. Some employees were exposed to phosphorus vapours over extended periods and developed a truly horrendous condition called 'phossy jaw'. A good description of the conditions under which these matches were made, and their consequences may be found in John Emsley's *History of Phosphorus*.[12] Matches no longer contain white phosphorus and will not ignite so readily as lucifers which is why they are sometimes called safety matches. In the 18th century phosphorus pieces were occasionally taken into coal mines to provide light in preference to candles to reduce the risk of explosion. Given the likelihood of the material bursting into flame and its highly poisonous nature, this would have been a hazardous activity. About the same time, luminous fish were also said to be used for the same purpose. This was a safer but no doubt antisocial remedy. In this case the luminosity was caused by luminescent bacteria colonising the decaying flesh.

Phosphorus has also found wide use in warfare since the early 20th century. At this time it was made on a large scale in electric furnaces. Burning phosphorus produces a dense white smoke of the oxide P_4O_{10}. The smoke has near ideal qualities as a screen and has been used effectively as such up to the present day. It has also been used as an incendiary device since WWI and had devastating consequences in the carpet bombing of European cities in WW2. Incendiary bombs and shells are still used in warfare but are banned against civilians by the Hague Convention.

During WW1, the tracer round was developed at Britain's Royal Laboratory. Rounds of ammunition were made with a hollow base into which white phosphorus and other incendiaries were packed. This allowed the flight of the round to be seen. The first rounds were made in 1915 where they were said to emit an erratic white trace for between 50 and 100 yards (45–90 m). By late 1916 improved versions were available. The Buckingham round had white phosphorus sealed within the bullet cavity. Access to the air was provided by a small hole filled with a low-melting point solder that melted upon firing. This allowed phosphorus to flow out and ignite. At the time, German zeppelins were carrying out regular bombing raids over London and proving difficult to destroy. In September of that year, Lt. W. L. Robinson mounted his Blériot Experimental 2c aircraft equipped with a Lewis gun. It was armed with the new tracer rounds and managed to destroy a zeppelin by igniting the hydrogen in its balloon. Subsequently other zeppelins were brought down and they were soon abandoned as a 'terror weapon' against civilians. Downing these dirigibles was hazardous for the pilots as the rounds had a short effective range.

Modern tracer rounds do not contain phosphorus but a range of other easily ignitable substances. Although they proved useful to the rifleman in directing fire, they also gave away firing positions to the enemy, so rounds were soon developed to provide a delayed trace or emit infrared radiation rather than visible light. The trajectory of these rounds differs from normal rounds as they lose mass during flight. The emission of light is the result of the heat within the gun chamber igniting the combustion mixture before it leaves the barrel. Tracer rounds can prove hazardous in other ways. In 2009 a training exercise near Marseille in France used tracer rounds that resulted in the accidental burning of 1200 hectares of dry shrubland and the evacuation of 1000 homes. There are also reports of tracer rounds smoking in aircraft when they were exposed to direct sunlight.

Antimony is an element closely related to phosphorus but is more metallic in nature. There are no reports of self-ignition but a peculiar form known as explosive antimony deserves mention. This is made by the electrolysis of a solution of antimony chloride using an antimony cathode. The resulting shiny black deposit of the element, if carefully washed and dried is highly reactive if rubbed. The material has been reported to rise to a temperature in excess of 100 °C. The reaction is unusual in being the result of a change of state. This involves a rearrangement of the atoms to form a more stable product and has been named β-antimony. The amount of heat liberated is in fact quite small but is enough to volatilise some of the impurities that are invariably present. Since the antimony is probably in a finely divided state, combustion of the material would appear to be possible. A sputtered film of non-crystalline germanium also crystallises explosively when the film is pricked with a pin. Germanium is in Group 14 of the Periodic Table but lies close to antimony in Group 15.

Bismuth also has some metallic properties. It is normally seen as groups of attractive crystals with a brilliant lustre and is often sold in mineral shops. A dark, finely divided form can be prepared by pyrolysing one of its compounds, bismuth mellitate. The resulting powder, sprinkled in the air, will catch fire to form the oxide. Compounds behaving in this way after heating are often referred to as pyrophores (see below). Mellitic acid used to prepare the compound is organic in nature and an unusual substance. It was first isolated from the rare mineral melite and is also found in coal. The molecule was discovered to have an unusually stable ring structure. It has been suggested that salts of mellitic acid may exist in Martian soils and might be an indicator of extraterrestrial life.

Sulphur belongs to the Group 15 elements and is included here on account of its flammability and low auto-ignition temperature of 230–240 °C. The element has been used for millennia as a fumigant and later as a component of explosive mixtures such as gunpowder. It is easily combustible and has been put to use in fire-lighting materials and this accounts for its earlier name of brimstone. In China matches known as 'light bearing slaves' contained sulphur and may have been used as early as the 10th century CE. Sulphur is not usually considered as spontaneously combustible although a slow oxidation in air can be detected at room temperature. As a fine dust, it has sometimes caused explosions.

2.6   CHEMICAL COMPOUNDS

2.6.1   The Hydrides

Most of the elements can be made to combine with hydrogen to form hydrides, some of the more familiar being water (H_2O) and the nitrogen hydride known as ammonia (NH_3). A few hydrides are spontaneously combustible but they are not thought to occur in nature with the exception of the phosphorus hydrides. These will be described later. The hydrides of carbon and sulphur along with the Group 17 halides, known as the halogen acids are well known compounds. Most of the remainder are somewhat obscure and rarely seen in the laboratory. The parent elements are mainly of low electronegativity and may be expected to be reactive toward the oxygen of the air.

No useful classification of the hydrides can be made but a gradual change in bonding type is observed across the Periodic Table. At the left of the table are the hydrides of Groups 1 and 2, the alkalis and alkaline earth metals. They correspond well with ionic compounds with the hydrogen atom accepting the outer electron of the metal. In the centre of the Table is the block of transition elements. In these elements, which are all metals, hydrogen atoms become trapped within the metal lattice and are given the name interstitial hydrides. Any bonding is metallic in nature. Two of the transition elements, osmium and iridium do not appear to have hydrides. Then to the right of the Table are the covalently bonded hydrides that include water and ammonia.

Beginning with the ionic hydrides in Groups 1 and 2, only barium hydride has been occasionally reported to self-ignite. Barium is the heaviest of the stable elements in these two groups and like its neighbour caesium in group 1, its outer electrons are well-shielded from the nucleus making it very reactive. These hydrides in common with others in these groups are solids at room temperature.

The transition element titanium forms a hydride that can be self-igniting – but only in a finely divided state. The only other member of this group is zinc hydride. This compound is moderately stable at room temperature but there have been reports of spontaneous combustion in the literature when it is in the presence of water. The hydrides of lanthanum and neodymium (lanthanoids) and uranium (actinoid) have also been reported to self-ignite.

Several hydrides of the non-metals have interesting properties, in particular those of boron and silicon. Boron resembles carbon in some respects but does not occur freely in nature. The hydrides can be made by heating magnesium metal with dry boric acid to form magnesium boride. By adding an acid to this compound, boron hydrides are formed.

The simplest of the boron hydrides is diborane whose molecular formula is B_2H_6. Since boron has a valency of three, the structure of diborane was at first inexplicable and it was not until the 1940s that it was realised that two of the hydrogen atoms were sharing their electrons with a pair of boron atoms.[13] These 'electron deficient' molecular bonds are rarely encountered. Diborane can be made by adding a strong acid to magnesium boride but it is impure and contains a mixture of several other boranes (Figure 2.6a). This compound is a gas at ambient temperature and spontaneously combusts in air to give a bright green flame. Upon oxidation boron oxide and water is produced. Diborane has one of the highest known heats of combustion.

The higher boranes have more complex structures. They have prefixes attached to their names according to the arrangement of their hydrogen atoms. There is tetraborane B_4H_{10} which has a spider-like shape in some orientations and is known as an arachno-form (Figure 2.6b) and there is nido-pentaborane-9, B_5H_9 so called because it has a basket-shaped ring structure (Figure 2.6c). The last is a liquid at room temperature and spontaneously combusts. In this compound the boron atoms have slightly

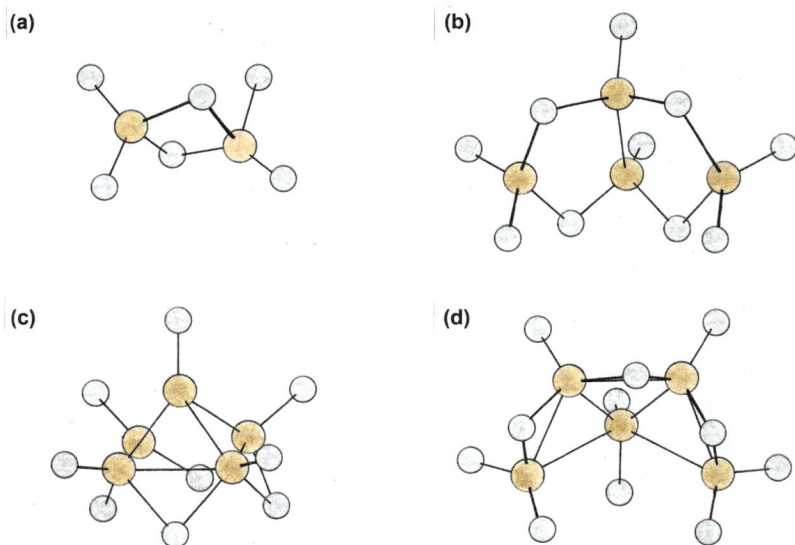

Figure 2.6   The structure of some boranes. (a) Diborane, (b) (arachno) tetraborane, (c) (nido) pentaborane (9), (d) (arachno) pentaborane (11).

different properties according to their positions. Arachno-penta-borane B_5H_{11} (= pentaborane-11) also self-ignites but is rapidly attacked by water (Figure 2.6d).

The terminal B–H bonds of these hydrides are weaker than the inner B–H–B bonds and therefore more likely to be broken, potentially leading to combustion although not all of the higher boranes self-ignite. The apical boron atoms found in some of these hydrides have high electron densities and these are open to attack by electrophiles – electron-deficient molecules or ions that readily accept electrons. Inner boron atoms with a lower electron density such as B–H–B have a lower electron density and are subject to attack by nucleophiles – molecules or ions that donate electrons. The overall reactivity of these remarkable compounds falls as their molecular weight increases.

The element silicon shares many of the properties of boron. This pair of elements form a diagonal relationship in the periodic table where similar properties are frequently recognised. Silicon hydride or silane as it is better known, also self-ignites. Discovered in the 1850s, this gas was first produced by the action of water on metal silicides. Silicides are compounds where silicon is united directly to a metal atom much in the same way that boron forms borides. Silane is most conveniently made from magnesium silicide. The silicide forms a black powder when an excess of magnesium powder is heated strongly with clean sand. Sand is composed mainly of silica (SiO_2) and the magnesium reacts with the oxygen of the sand to form magnesium oxide. Then more magnesium combines with the resulting silicon to form the silicide (Mg_2Si). Silane (SiH_4) is liberated upon addition of a strong acid to the silicide. If the gas is allowed to bubble through water into the air, it immediately catches fire, often producing small explosions accompanied by white fumes of silicon dioxide (Figure 2.7). A mixture containing 1.5% silane in nitrogen gas will combust in an atmosphere containing as little as 0.1% oxygen. The combustion however can only be detected by analysing the products in the reaction vessel since no flames can be seen. Experiments with such mixtures suggest that ignition is determined by the ratio of silane to oxygen molecules rather than their concentration. It has led to the suggestion that complex competing reactions are taking place.

If silane is placed in contact with air-saturated water, it slowly decomposes, becomes foggy and no longer combusts

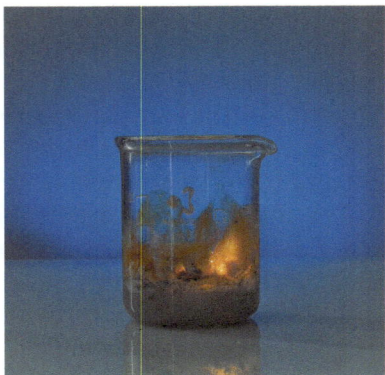

Figure 2.7   The spontaneous combustion of silane showing fumes of silica (SiO_2). © Declan Fleming, Alamy Stock Photo.

spontaneously. Pure silane is used on a large scale by the semiconductor industry to produce deposits of silicon for integrated circuits and this has sparked considerable interest in its properties. There are in fact several silicon hydrides and all are unstable.

2.6.2   Oxides

Elements such as sodium produce just one series of compounds with the halogens. The most familiar example is sodium chloride or common salt, NaCl. Other elements behave differently and iron is a good example. Iron belongs to the transition elements where electrons can be promoted from an inner shell to the outer valency shell. It forms two chlorides namely iron(II) chloride, $FeCl_2$ and iron(III) chloride, $FeCl_3$. In the first case, the outer two electrons of iron are attracted towards the two chlorine atoms which are strongly electronegative. In the case of iron(III) chloride, an electron from the inner d shell of iron is promoted and used as an additional valency electron to form iron(III) chloride. In the presence of oxygen, iron(III) is the more stable because oxygen in the atmosphere provides an environment that encourages electrons to be withdrawn from the iron. In the absence of oxygen, in more reducing environments, iron(II) is more stable.

Iron has already been shown to be spontaneously combustible when finely divided. When such iron is prepared from the tartrate salt however, it is now believed that it is iron(II) oxide and not metallic iron that self-ignites. The preparation of this form of iron involves heating iron(II) tartrate in a pyrex test tube until it blackens. The tube is then cooled and the contents are knocked out into the air. An impressive display of bright orange sparks results. Iron(II) oxide (FeO) is known to be highly reactive towards oxygen, forming the higher, familiar iron(III) oxide.

Other iron compounds may also show similar behaviour. For example, the magazine *Popular Science* described in 1933 a deposit that formed on iron hoops placed around vinegar barrels. It ignited on the application of a match. This theme of converting a lower (*e.g.* iron(II)) oxide to a higher (iron(III)) oxide also appears to operate in the case of tin and also perhaps chromium. The lower oxide of tin, known as tin(II) oxide, has been observed to smoulder when exposed to air. It is formed by heating tin(II) oxalate. The final product is tin(IV) oxide, SnO_2. The case for lead is interesting and in some respects similar. It too forms two simple oxides, lead(II) oxide, (PbO) and lead(IV) oxide (PbO_2). Heating lead(II) tartrate in the absence of air would be expected to result in the formation of the lower oxide, since tartrates and oxalates undergo similar reactions. The resulting material self-ignites but the process is not fully understood. It will be discussed further below. Yet another example is shown by cerium(III) oxide. This compound is easily oxidised to cerium(IV) oxide, CeO_2.

The compound popularly known as quicklime or calcium oxide reacts violently with water liberating much heat, a process known as slaking. This is a different type of chemical reaction and here changes in oxidation state do not occur. Furthermore, the material itself does not combust but provides sufficient heat to activate the combustion of other flammables. Quicklime is made in large kilns from limestone, a form of calcium carbonate, which gives up its carbon dioxide on heating to yield the oxide, a process known as 'lime burning'. Calcium belongs to the Group 2 elements along with magnesium, strontium and barium. There is a progressive increase in the heat of formation of the hydroxide from the oxide as the atomic weights of these elements increase. The conversion takes place on the addition of water. Strontium and barium both have higher atomic numbers than calcium and the heats of formation of their hydroxides are even greater. However, these elements are much less common than calcium. Strontium carbonate is difficult to convert to the oxide by 'burning' and this cannot be done at all with barium. These oxides have to be made using other methods.

The slaking of lime has received much study as it is a major industrial chemical. The process has been shown to be accompanied by a temperature rise of up to 468 °C where light emission may be observed. However, the slaking process is difficult to reproduce. It has been found that the temperature of the lime burning operation, its duration, the nature of the raw material and the particle size of the product all influence the rate of heat production. The thermal conductivity of quicklime is also low which helps to maintain a high temperature within the mass. Another interesting feature of slaking is the almost doubling in volume of the product, calcium hydroxide. Knowledge of this has sometimes been used to split rocks in quarrying operations. The slaking process is not spontaneous combustion since no oxidation is occurring although the reaction can be made to initiate combustion in other materials.

Since quicklime is made by heating limestone, a common sedimentary rock, the material has been known for thousands of years. Theophrastus (4th century BCE) recounted that ships laden with new togas were known to burst into flame owing to water splashing on the material, as they were bleached with a mixture of lime and sulphur. The properties of quicklime were described by the Roman authors Livy (*ca.* 60 BCE–15 CE) and Quintus Curtius (1st century CE). In the time of Alexander VI *ca.* 230 CE, a mixture of quicklime and asphalt rolled into a ball was called 'automatic fire'. On contact with water, it burst into flame. The *pyr automaton* is based upon a mixture of sulphur, resin, quicklime and naphtha. Ignition also occurs upon water contact.

A work attributed to Julius Africanus (*ca.* 170 CE) gave a recipe that included sulphur, common salt, resin, charcoal, asphalt and quicklime mixed into a paste that could be smeared on the woodwork of an enemy siege engine at night. Once the morning dew arrived, the mix would ignite and destroy the engine. A thick fluid called *nephthar* was also known to the ancients. It could cause timber to combust and is thought to have been a mixture of quicklime with either sulphur, petroleum or a mixture of the two. The reaction of quicklime with water would have raised the temperature sufficiently for combustion to take place.

In the Middle East, where natural seepages of petroleum occur, it could not have been long before mixtures of quicklime and petroleum were found to be ignited by water. Sulphur was also known to the ancients, and owing to its low combustion temperature was easy to ignite. Bags containing sulphur and quicklime were said to have destroyed a fleet of Chin vessels on the Yangtze River, China, in 1161 CE. Early pyrotechnic mixtures are also known, such as powdered tin and quicklime, again ignited by water.

2.6.3 Azides, Nitrides and Nitrates

About 1400 tonnes of the explosive TNT (trinitrotoluene) resides in the wreck of the USS Richard Montgomery off the coast of southern England. It is well known to the local inhabitants as it is widely believed that it may explode without warning. The ship arrived in August 1944 with a large cargo of bombs, and mistakenly guided into a berth that was too shallow and it ran aground. All efforts to move it failed and to make matters worse it broke in two in September of the same year. Although some explosives had been removed before the break, over 13 000 pieces of ordnance remained, their removal having been halted owing to cost. Ammonium nitrate was sometimes added to TNT and this is cause for concern. Wet ammonium nitrate has been known to react explosively with iron, so a seawater leak into a bomb casing could be disastrous. The Ministry of Defence had assumed that all the bombs were unfused, but this was later disproved. The fuses contained lead azide, a compound of lead and nitrogen that explodes violently on impact. Fears have arisen because if water penetrates the detonator, it could complete a reaction between the lead azide and the copper in the detonating cap. The result would be the formation of copper azide which is even less stable and could explode during a movement of the wreck. The azides are a group of compounds where metals and some organic compounds are combined directly with an N_3 nitrogen group. Many azides are unstable and explosive at room temperature.

Ammonium nitrate has achieved notoriety as a peacetime explosive that has ended thousands of lives. It is used as a fertiliser as it is an effective nitrogen source for crops. The trouble began in 1916 when over a tonne of the chemical exploded whilst being evaporated from the solution at Oakdale, Pennsylvania. Three years later there was an enormous explosion of the chemical at Faversham in Kent when 115 people were killed. One of the worst incidents was at Oppau in Germany where approximately 200 tonnes of the nitrate exploded in a silo. The nitrate was mixed with ammonium sulphate but after exposure to the moist atmosphere it had caked making it difficult to remove. A tried and tested method of breaking up the fertiliser at Oppau was to use a small charge of dynamite. On this occasion there was a huge explosion that killed over 500 people. The cause was assumed to be the detonation by dynamite but nobody survived to tell.

There have been many incidents since that time. Evidence suggests that some, perhaps most, were the result of overheating by accidental fires that could have been prevented. Other incidents however are more worrying and suggest that spontaneous ignition may have taken place. The Oakdale incident mentioned above may have been caused by the presence of impurities. An incident at Cartagena, Spain, in 2003 was thought to have been the result of a self-sustained decomposition of the nitrate but in this case it did not end in an explosion. Self-sustained decomposition, or SSD, occurs when a small volume of material starts to decompose and then spreads rapidly throughout the mass.[14] If the mass is substantial, the rate of loss of heating resulting from the exothermic reactions through conduction and convection cannot keep up and a 'thermal runaway' occurs. At a certain point, a deflagration to detonation transition is reached as the expanding gases become supersonic. Self-sustained decomposition can be caused by the presence of hot surfaces or may occur spontaneously. The impure salt appears to be most at risk and there have been incidents with NPK fertilisers where the nitrate is mixed with phosphates and potassium salts.[15] In one case the nitrate was contaminated with oil and it was suggested that the resulting explosion was caused by the shock of a falling object. Experiments have also shown that organic matter can cause decomposition of the nitrate as low as 100 °C. Catalytic action may also be involved.

2.6.4 Other Compounds and Mixtures

A wide range of inorganic compounds have exhibited self-ignition behaviour, several of which contain the elements phosphorus or sulphur. These include thorium oxysulphide, lead sulphide (at 40 °C), some sulphides of the alkali metals and sodium hypophosphite. A few selenium compounds such as uranium diselenide are also known to flame, selenium being closely related to sulphur. Some carbides such as that of iron self-combust. Many carbides produce the gas acetylene (C_2H_2) by contact with water and this has also been noted to spontaneously ignite on occasion. There are also some more unlikely candidates such as titanium chloride which can ignite if wetted.

Among the vast array of organometallic compounds a good number either self-ignite or are easily combustible. In these compounds, a metal atom is linked directly to a carbon atom. The carbon will also be bonded to other carbon and hydrogen atoms. One of the best-known is zinc diethyl. This compound which exists as a liquid at room temperature was first prepared in 1848 by Edward Frankland. Edward became interested in chemistry as a schoolboy after attending a trial where hydrochloric acid was allowed to escape from a Liverpool chemical works and cause pollution. At the suggestion of the German scientist Robert Bunsen he investigated the constitution of some common organic compounds. In his search for the nature of 'ethyl' he recognised that the iodine in ethyl iodide could probably be removed using a reactive metal such as zinc. It was during these experiments made in Bunsen's laboratory that he prepared the first organometallic compound, zinc diethyl. It proved to be an important experiment since from it, he later propounded the theory of valency, one of the foundation stones of chemistry. Zinc diethyl combustion is usually demonstrated by squirting a syringe full of the liquid onto an inert surface in a fume cupboard. This is necessary since on ignition, dense fumes of zinc oxide are produced. The compound also reacts violently with water. The oxidation involves the formation of peroxides and these reactive molecules may be involved in the ignition process.

The paper found in older books is acidic mainly because alum (aluminium potassium sulphate) was added to the pulp as a strengthener and mordant. Unfortunately this leads to the catalytic oxidation of the cellulose by the air. It eventually makes the paper brittle and crumbly, a process known in the trade as 'slow fire'. A few years ago attempts were made to remove this acidity by subjecting them to zinc diethyl. Zinc oxide is a base and is formed during zinc diethyl decomposition. As a liquid the diethyl should easily penetrate the paper of books and react with the acidity present in paper, neutralising it, without causing discolouration. Unfortunately a series of explosions took place in the trials so the method had to be abandoned.

Spontaneously combusting organometallics are found in four of the Group 1 elements, lithium, sodium, potassium and rubidium. Most organo-lithium compounds, such as methyl lithium are known to self-ignite.[16] This molecule exhibits strong covalency owing to the small size of the lithium atom, where the seat of the reactivity lies. Aliphatic compounds containing sodium, potassium and rubidium are even more unstable and many self-combusting examples are known.

The transition elements copper and silver also form unstable organometallics. Methyl copper and methyl silver are particularly reactive and may explode upon drying. In Group 13, trimethyl aluminium is rapidly oxidised in air. It is structurally similar to diborane and like it, possesses electron deficient bonds. Indium and thallium, belonging to the same group, form liquid trimethyls which spontaneously ignite.

Nitrocellulose, or 'celluloid' is not an organometallic compound but it has had a violent history. In an attempt to replace ivory as a material for billiard balls in the 1800s, balls were prepared using camphorised nitrocellulose. This had the alarming property of occasionally exploding when the balls collided.

Nitrocellulose was once employed to make cinematographic film. Unfortunately it is highly flammable and can self-combust. A major fire, accompanied by explosions, occurred in the 20th Century-Fox storage site in New Jersey in 1937. This was the main repository of the Fox Film Corporation holding material from the silent film era up to 1932. A heat wave was occurring at the time and is believed to have been a factor, with air temperatures exceeding 38 °C. The films were stored in containers within confined areas with little ventilation. It is thought that the nitrate began to decompose releasing flammable gases that subsequently ignited. Eyewitnesses reported flames up to 30 m long blowing horizontally out of the windows. Many films, known from just a single copy, were destroyed.

At the port of Tianjin, China nitrocellulose stored in a warehouse spontaneously combusted on a warm day and detonated about 800 tonnes of ammonium nitrate stored nearby. After a number of tragic events, including fires in cinemas nitrocellulose has now been replaced by the less flammable cellulose acetate.

Several chemical mixtures have been used to demonstrate self-ignition in school science classes. In one, potassium permanganate is added to an alcohol such as glycerol. This is an example of a mixture of a strong but reasonably stable oxidising agent, permanganate reacting with a carbon-containing fuel. Held in an unmixed state, the mixture has been used for fire-making in survival kits as the products are non-toxic. Permanganate salts are well known for their fire-assisting properties as they contain a large proportion of oxygen. Alcohol-soaked paper held above a mixture of potassium permanganate and sulphuric acid also ignites. The alcohol is thought to burn due to the production of ozone, a particularly reactive form of oxygen. This is a dangerous experiment and best left to the imagination. Another is a mixture of superglue and cotton wool. The glue contains cyanoacrylate, a compound which reacts with hydroxy (OH^-) groups in the wool. The temperature rises to the ignition point of wool in a few minutes. This is another example of an exothermic reaction not involving molecular oxygen. Substances that catch fire upon mixing are termed hypergolics and some have found use in rocket engines. A particularly effective mix is that of hydrazine and nitrogen tetroxide.

2.7 THE PYROPHORES

This term has been applied generally to any material that combusts spontaneously in air but in this section it is confined to those substances that have been heated strongly prior to their action. One example, bismuth formed from the mellitate has already been noted.

The first pyrophoric material is believed to have been discovered by Wilhelm Homberg (1652–1715). He obtained it by distilling excrement with alum in an attempt to produce an 'odourless white oil' for use in alchemy. On breaking open the tube, the roasted residue burst into flame.[17] After a lapse of nearly 30 years, Homberg re-visited his experiment and concluded that the phenomenon was the result of atmospheric moisture reacting with the roasted salt. This caused a rise in temperature which ignited the oil. The observation recalls the reaction taking place when quicklime is slaked. He called his mixture a 'pyrophorus', a term later to be applied to many other spontaneously combusting solids. More work soon followed and it became clear that other organic materials could be used to produce the same effect when roasted with alum. These include wood, flour and even honey. One of Homberg's colleagues, Louis Lemery suggested that the sulphate in the alum was being reduced to sulphur and the latter, being combustible, was responsible for the ignition. Further research revealed that the alum could be replaced by other sulphates such that of sodium so it was clearly a reaction (or reactions) between a sulphate and organic matter. A significant observation was made by the Swedish chemist Carl Scheele who suggested that an alkali sulphide was responsible for the combustion. He went on to demonstrate pyrophory by heating potassium sulphide with charcoal. The chemist Lavoisier also took an interest and in the process of experimenting with pyrophors discovered the gas carbon monoxide. The chemist William Jensen reviewed this work in 1989. Through many lines of enquiry, Jensen concluded that during the roasting process, potassium sulphate of the alum was reduced to potassium sulphide. This was caused by the carbon present in the organic matter reducing the sulphate and producing the gases carbon monoxide and dioxide. Potassium sulphide in a finely divided state is rapidly oxidised by the air back into the sulphate and the resulting heat appears sufficient to ignite the carbon. It would seem that other sulphates containing an alkali metal should behave similarly, although there appears to have been no comprehensive study. Other pyrophores were later discovered which did not contain sulphates. One in particular is that formed from the pigment Prussian Blue.

This pyrophore was first described by an American chemist, R. Hare and apparently brought into production as 'Hare's pyrophorus'.[18] It was said to have been sprinkled onto tobacco to light pipes in the 19th century. This pyrophore however had already been known for some time. The pigment Prussian Blue was discovered by accident in the early 1700s. Curiously, it was obtained during an experiment to produce another pigment called carmine which was obtained from the wing cases of the cochineal scale insect. Carmine was an important and expensive dye at the time and experiments were performed to improve its colour-fastness and intensity. It was probably during such an experiment that cochineal was heated with potassium sulphate but instead of becoming deep red, the product became dark blue. This may have been the result of contamination of the potassium sulphate with iron. Needless to say, the contamination caused much confusion since sulphates, as noted above, could also induce pyrophory. The use of impure chemicals was one of the main reasons for the slow progress of chemistry prior to the 19th century.

It was soon discovered that one of the two key ingredients of Prussian Blue was iron and the compound is now known as iron(III) ferrocyanide. Although it contains cyanide, it is not poisonous as the cyanide is strongly bound to the iron. Prussian Blue is an intensely blue pigment and is the result of the molecule containing iron in two oxidation states, iron(II) and iron(III). It is termed a mixed valency compound. The discovery of its pyrophoric action probably came about early in the 19th century since the compound is normally produced as a water suspension. This then needs to be dried down into a powder for use. The pigment starts to decompose around 250 °C with the loss of water. Further roasting causes it to become pyrophoric (Figure 2.8). The roasting of Prussian Blue was thought to yield iron in a pyrophoric state, but it is more likely to be a finely divided form of iron(II) oxide and can be compared with the iron(II) tartrate procedure noted above. It is clearly quite a different process to that occurring in the roasting of sulphates with organic matter since there is no sulphur in the Prussian Blue molecule.

Pyrophoric lead also remains something of an enigma. It is made by dissolving a lead compound in tartaric acid. Being an organic acid, it is assumed that the tartrate on heating is decomposed, with the release of carbon, accompanied by oxides of carbon, water and perhaps hydrogen. Some of these products could reduce any lead oxide present to a highly divided form of lead metal which may then catch fire. This hypothesis was

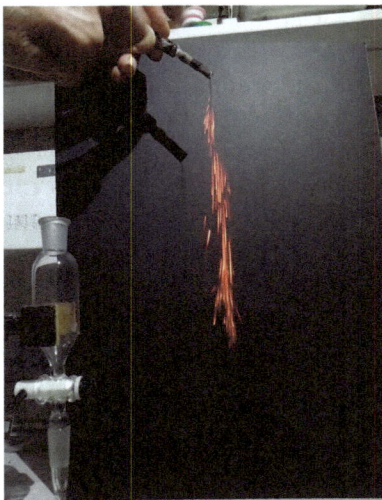

Figure 2.8   Pyrophoric iron. The pyrophore is being tapped out of a small ampoule when it immediately catches fire.

suggested by W. van Rijn in 1908 but does not appear to have been followed up.

There is an interesting account of pyrophoric lead in the *Boy's Book of Metals* by John Henry Pepper first published in 1861.[19] He mixed the calculated amount of lead acetate with ammonium tartrate to obtain lead tartrate as a white precipitate. For his demonstrations, Pepper used a test tube drawn out into a capillary at one end and containing about one gram of the tartrate. This was heated over a gas burner until the substance stopped smoking and was then allowed to cool. When tapped out into the air, the dark powder bursts into flame. Later experiments showed that an open tube worked equally well provided the demonstration was completed quickly (see Box 2.2).

BOX 2.2   EXPERIMENTS WITH LEAD TARTRATE

Using the above technique, the writer subjected samples of lead tartrate to heat in a muffle furnace at a range of temperatures. When heated for two hours there was no loss in mass or change in colour below 225 °C. Above this temperature mass loss was rapid and the colour changed to a dark

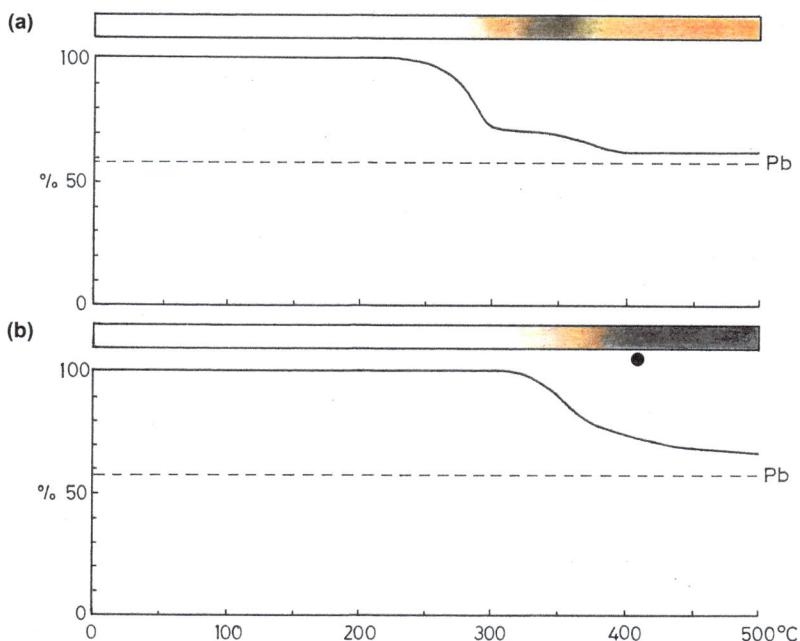

Figure 2.9 Thermal decomposition of lead tartrate. (a) Samples held at a range of temperatures for 2 h. (b) Held at temperature for 3 min. Solid lines show the percentage change in mass. Broken lines show the lead content as a percentage lead tartrate by mass. The black spot indicates the point at which spontaneous combustion began.

chocolate brown up to 300 °C (Figure 2.9). Between 300 and 360 °C the material blackened with a further loss in mass. Between 350 and 500 °C the powder changed to the orange–yellow colour of the mixed oxide known as red lead Pb_3O_4. Heating for this period of time provided opportunity for oxygen in the air to react with the products of decomposition and spontaneous combustion was not observed.

Heating the tartrate rapidly for three minutes at a range of temperatures produced a different result. Here decomposition and discolouration occurred at 320 °C, about 100 °C higher than before. The colour change also differed although there was still a rapid mass loss with blackening setting in at 375 °C. Red lead was not produced as there was insufficient opportunity for oxygen to react with the product. The colour change from pale brown to black probably corresponded to the formation of pyrolysis products such as caramelins, finally resulting in black particles of carbon. The black mass contained about 90% lead by weight. Spontaneous combustion of this material began when the temperature reached 415 °C. The lead will be formed in a finely divided state as it is prevented from coalescing into metallic globules by the carbon. The experiments suggested that the pyrophoric material contained both lead(II) oxide and some metallic lead.

The structure of the tartrate (Figure 2.10) shows that the lead atom is bonded directly to oxygen making the initial formation of oxide likely. The oxidation of both metallic lead and this lower oxide are both exothermic reactions leading to a rise in temperature. Both could behave as pyrophors. It is also noteworthy that lead(II) sulphide, PbS, has also been known to combust at temperatures as low as 40 °C.

In the light of this study, tartrates were prepared of elements of a similar electronegativity to lead. Praseodymium tartrate was subjected to the same heating procedure and the residue self-ignited in the same way. The tartrate salt of copper is of interest given the tell-tale green flame produced by compounds of this element. The tartrate residue again combusted but only the reddish yellow sparks of burning carbon were apparent. Clearly there is more to be learned about the phenomenon.

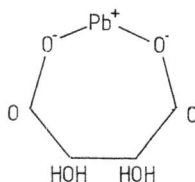

Figure 2.10   The structure of lead tartrate.

FURTHER READING

N. N. Greenwood and A. Earnshaw, *Chemistry of the Elements*, Elsevier, Amsterdam, 2nd edn, 1977.

J. C. Kotz, P. M. Treichel and P. A. Harman, *Chemistry and Chemical Reactivity*, Thomson Learning, Chicago, Ill, 11th edn, 2023.

J. W. Mellor, *Comprehensive Treatise on Inorganic and Theoretical Chemistry*. Longman, London, vol. 21, 1917.

J. R. Partington, *A History of Greek fire and Gunpowder*, Johns Hopkins Press, Baltimore, Mayland, 1960.

S. J. Pyne, *Fire. A Brief History*, University of Washington Press, Seattle, 2001.

REFERENCES

1. D. A. Frank-Kamenetskii, *Diffusion and heat transfer in chemical kinetics*, Plenum, New York, 1969.
2. A. S. H. Makhlouk and A. Barhoum, *Fundamentals of Nanoparticles*, Elsevier, Amsterdam, 2018, DOI: 10/1016/C216-0-01899-5
3. J. C. Bay, Jean Senebier, *Plant Physiol.*, 1931, **6**, 188–193.
4. H. Cavendish, Experiments on Air, *Phil. Trans. Roy. Soc.*, 1785, **15**, 372–384.
5. R. A. Berner, GEOCARBSULF: a combined model for Phanerozoic atmospheric O_2 and CO_2, *Geochim. Cosmochim. Acta*, 2006, **70**, 5653–5664.
6. C. M. Belcher and J. C. McElwain, Limits for combustion in low O_2 redefine paleoatmospheric predictions for the Mesozoic, *Science*, 2008, **321**, 1197–1280.
7. P. G. Falkowski, M. E. Katz, A. J. Milligan, K. Fennel, B. S. Cramer, M. P. Aubry, R. A. Berner, M. J. Novacek and W. M. Zapol, The rise of oxygen over the past 205 million years and the evolution of large placental mammals, *Science*, 2005, **309**, 2202–2204.
8. A. J. Watson and J. E. Lovelock, in The dependence of flame spread and probability of ignition on atmospheric oxygen: an experimental investigation, in *Fire phenomena and the Earth System: an interdisciplinary guide to Fire Science*, J. Wiley, New York, 2013, pp. 273–287.
9. B. B. Stephens and P. S. Bakwin, *et al.*, Application of a differential fuel-cell analyser for measuring atmospheric oxygen variations, *J. Atmos. Ocean Technol*, 2007, **24**, 82–94.

10. M. Mitu, E. Brandes and W. Hirsch, Ignition temperatures of combustible liquids with increased oxygen content in the CO_2 + N_2 mixture, *J. Loss Prevent. Proc. Ind.*, 2019, **62**, DOI: 10.1016/j.jlp.2019.103971.
11. M. Crosland, *The Society of Arcueil, A view of French Science at the time of Napleon I*, Heinemann, London, 1967.
12. J. Emsley, *The shocking history of phosphorus: a biography of the Devil's element*, Cambridge University Press, 2000.
13. M. G. Davidson, K. Wade, T. B. Marder and A. K. Hughes, *Contemporary Boron Chemistry*, Royal Society of Chemistry Special Publication, 2000, DOI: 10.1039/9781847550644.
14. J. Akhavan, *The Chemistry of Explosives*, Royal Society of Chemistry, Cambridge, 3rd edn, 2011.
15. R. Hadden and G. Rein, Small-scale experiments of self-sustaining decomposition of NPK fertilizer and application to the events about the *Ostedijk* in 2007, *J. Hazard. Mater.*, 2011, **186**, 731–737.
16. *Organometallic compounds: Synthesis, Reactions, and Applications*, ed. D. K. Verma and J. Aslam, Wiley-VCH, New Jersey, 2023, DOI: 10.1002/9783527840946
17. W. B. Jensen, Whatever happened to Homberg's pyrophorus?, *Bull. Hist. Chem.*, 1989, **3**, 21–23.
18. R. Hare, New Pyrophorus, *Am. J. Sci.*, 1831, **19**, 173.
19. J. H. Pepper, *The Boy's Book of Metals*, G. Routledge, London, 1875.

Spontaneous Combustion in the Hands of Nature

3.1 INTRODUCTION

In the previous chapter we have seen how chemists have succeeded in uncovering a wide range of materials capable of self-ignition. In the laboratory, conditions can be controlled, experiments undertaken and observations made. In the natural environment, the situation is different. Fire initiation usually occurs without observation. For example, a forest fire may start by a fragment of glass acting as a lens concentrating sunlight onto tinder-dry brushwood. This would not be obvious in the aftermath of the fire and it could be attributed instead to a self-ignition event. Many cases of this type will be found in this chapter. In the previous chapter a low temperature limit was placed on spontaneous combustion events. In this chapter there is in general less information available on the temperature at which combustion begins and a higher temperature limit of *ca.* 250 °C is more realistic.

We begin by looking at physical processes capable of concentrating the sun's rays sufficiently to exceed the auto-ignition temperature of natural materials. Next an unusual example is drawn from the geological literature. This involves a physical

Luminous Phenomena: A Story of Spontaneous Combustion, Phosphorescence and Other Cold Lights
By Allan Pentecost
© Allan Pentecost 2025
Published by the Royal Society of Chemistry, www.rsc.org

process that might not have involved ignition *per se*. There follows a section on fires resulting largely upon biological processes. It includes those occurring in harvested organic matter such as hay and manufactured materials such as flour and cloth. We conclude by looking at fires associated with living things both animals and vegetables. It will be clear that in all of these cases, the scientist is faced with fragmentary data that is not amenable to controlled conditions. Often, the science will become part of a wider forensic investigation.

3.2 SOME PHYSICAL PROCESSES

Ice can be shaped easily using simple tools and when formed into a biconvex lens can be made to set fire to tinder. The ice needs to be clear otherwise the light is scattered in too many directions and will not focus sufficiently. Even a crudely formed lens about 4 cm thick and 15 cm wide can achieve this and they are often demonstrated in survival manuals. The ice may also be formed into a sphere but the focal length is reduced to about 2 cm for a 10 cm diameter sphere. Ice sculptors have made spheres to 1 m wide or more but it becomes challenging to obtain a clear piece of these dimensions. Exceptionally clear ice is sometimes seen in icicles and it is possible that these too could start fires if flammable material is in their vicinity. Clear quartz spheres produce the same effect as ice and there are reports of fires being started by them when left upon linen in sunlight.

Lenses manufactured to cause ignition have been used since ancient times. Pliny the Elder mentioned glass vessels filled with water used to cauterise wounds. A burning mirror was placed in the temple of the Vestal Virgins presumably to re-kindle the 'eternal' Vestal flames. The basic physics of these devices was understood by the Chinese as early as the 11th century CE.

In 2013 a UK newspaper reported the near destruction of a car that was parked close to the skyscraper at 20 Fenchurch Street, London. The structure, known locally as the 'Walkie-talkie' has a concave façade on the south side and is constructed largely of glass. Concave surfaces are well known for their ability to focus the sun's rays onto small areas resulting in the development of high temperatures. The car suffered damaged panels and mirrors caused by the melting and warping of plastic components.

Nearby shopkeepers had complained of carpets catching fire and of smouldering front doors. A similar structure existed at the Vdara Hotel, Las Vegas and became known as the 'Vdara Death Ray'. Another example is the world's largest solar furnace situated in the French Pyrenees. Temperatures exceeding 3000 °C have been achieved with the device.

Archimedes used the burnished shields of soldiers to create a large curved mirror. It was said to have been focussed onto the wooden ships of Claudius Marcellus' fleet resulting in the destruction of at least one of them. In 515 CE a similar tactic was employed by the philosopher Proklos to destroy ships sent by Vitrulinus against Emporer Anastasios. These well-known accounts intrigued later philosophers and in 1747, Count Buffon of the Natural History Museum of Paris decided to test the idea. He succeeded in burning a plank of wood by focussing the sun's rays with mirrors at a distance of 45 metres. More recently, I. Sakkas from Greece did a similar experiment using 60 sailors with polished shields. They managed to burn a wooden ship at a distance of 50 m. More recent experiments have met with less success and it appears that unless the target can be kept absolutely still, ignition is unlikely to occur.

3.2.1 Drops of Water

Given the fine state of division of some plant materials such as cotton bolls, cotton grass and the amadou obtained from fungi, we might expect that localised heating would lead to their combustion. Focussing the sun's rays through clear droplets of water or resin might do the trick (Figure 3.1). Gardeners in the UK sometimes explain that the reason crops should not be

Figure 3.1 Raindrop on a pine needle. Drop about 2 mm in diameter.

watered at midday is to prevent damage caused to plant surfaces by sunlit water drops. Spherical droplets act as lenses but light from the sun will be focused very close to the droplet surface. The focal distance from the centre of a sphere is a function of its diameter and the refractive index of the material it is made from. A spherical water drop of diameter 2 mm whose refractive index is 1.33 gives a focal length of 2.02 mm. In other words the focal point for the sun's rays is close to 1 mm from the drop surface. Any object at this distance would be subject to high humidity that could retard ignition, although as drop size increases, the focal point recedes from the surface. Drop size however does not increase indefinitely as the surface tension forces holding the drop vertical are soon overcome by gravity.

A recent study suggests that high temperatures could in fact be generated by water drops.[1] The researchers photographed droplets on vegetation and by this means were able to describe their form mathematically. They then calculated light paths through the drops at a range of solar angles and concluded that some droplets could concentrate sunlight up to 178 times. For small spheres this increased to 700 times. It was concluded that for water-repellent leaves, droplets had the potential to start fires if small waxy hairs were present on the leaves. Further experiments placing small glass beads and water drops over maple leaves demonstrated local damage by sunburn but no fires were reported. The highest temperatures were obtained with solar elevations exceeding 30°. The water-repellent leaves of some aquatic plants showed the greatest promise but the temperature was insufficient to start a fire.

The author carried out some further experiments with the simple apparatus described in Figure 3.2. He focused sunlight through a water drop onto a blackened surface upon which a range of organic compounds were thinly spread. These compounds had known melting points. The results indicated that at an air temperature of 22 °C, a solar elevation of 52 degrees, and a water drop about 2 mm in diameter, the surface attained a temperature between 53 and 81 °C. With a vertical sun, and a higher air temperature, greater values might be achieved. However it appears unlikely that the temperature could get much higher than this, particularly when it is considered that a vertical sun passing through a water drop would rarely impinge upon a

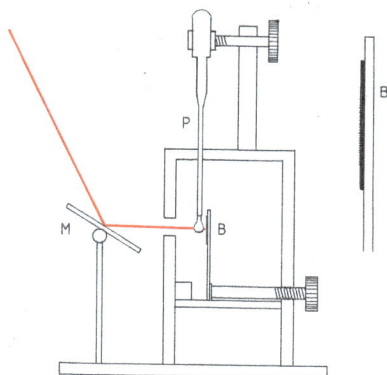

Figure 3.2 Apparatus for measuring surface temperatures of sunlight passing through water drops. Sunlight (red line) is guided by the
plane mirror M onto a water drop suspended from a pipette P.
Behind the drop, a black surface, B is covered with a thin layer of
the test material and the distance is adjusted by the screw.

dry surface. Larger drops can also be produced. Water con-
densing on a plane sheet of glass can produce drops up to 7 mm
across.

Again, larger drops should generate more heat but these drops
are distended by gravity into pear shapes. They would be unlikely
to remain stable for long periods unless conditions were exceptionally calm. The nature of the surface being irradiated also
needs consideration. If the heated area is small in size the heat
would be rapidly radiated and conducted away. Some would also
be reflected. The geometry and nature of the fuel would also need
to be favourable. For example, a black heat-absorbing surface
containing flammable volatiles would enhance solar heating.
Particular orientations, and the presence of deep depressions
where the heat may be concentrated and trapped, forming hot
spots, could be especially vulnerable to ignition (Figure 3.3).

While the heat provided by sunlight has been seen to have
potentially damaging effects through its effect on dry vegetation
it has also proved useful as a firelighter for cooking food and
keeping us warm. Ultimately this source of heat comes from
nuclear reactions in the sun's core. It came as a surprise to
geologists when they discovered evidence of a nuclear reaction
close to the earth's surface as will be described in the next
section.

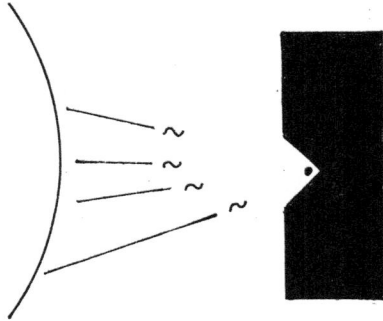

Figure 3.3   The sun's rays focussed by a water drop (right) onto a dark
depression (left) containing flammable material.

3.2.2   Oklo Mine

The search for uranium ores occurred late in industrialisation as
its value as a nuclear fuel was not recognised until well into the
20th century. Natural uranium occurs as several isotopes. They
include U^{238} the most abundant isotope (>99% by weight) and
U^{235}, of value in nuclear reactors, close to 0.71%. These proportions have been found to vary little from region to region.
Worldwide searches were made for uranium ore including the
Gabon where uranium oxide was discovered at the Oklo Mine.[2]
Here, geochemists discovered that the isotope ratio of the
uranium differed significantly from all previous measurements.
It led to the suggestion that a natural nuclear reaction had taken
place in the distant past. Further work showed that uranium
fission products were present and several sites were identified.
Nuclear fission reactions require moderators and in the case of
Oklo this is believed to have been supplied by groundwater. As
the nuclear reaction commenced, the ground would have heated
up, evaporating the water. The reaction would then have stopped
until more water seeped into restart the process. This seems to
have occurred over a period of about a million years, generating
around 100 kilowatts of energy. Although small in scale, the reaction would be sufficient to make the rock hot. Since the reactors operated about 2 billion years ago, there is little chance
that any evidence of this heating would have survived. At about
the same time, the earth's atmosphere became oxygen-rich in the
'great oxidation event' caused by photosynthesis. Uranium only

becomes mobile in oxygen-rich water and it is possible that it became sufficiently concentrated in the rocks to allow the reaction to take place.

The Oklo event was unusual but hot rock is common at the earth's surface in the vicinity of volcanoes and hot springs. Combustion of vegetation is well known in the vicinity of flowing lavas and ash clouds known as *nuées ardentes* where temperatures may exceed 500 °C. Hot rock also arrives from space in the form of meteorites, where the material becomes heated by passage through the atmosphere. Large meteorites impact the earth at high speed and further heat is generated on collision. Many meteorites also contain phosphides that are potentially self-combustible. While these phenomena, together with the heat generated by lightning strikes, can all lead to ignition of combustibles, this chapter is more concerned with little researched, near-ambient examples of self-ignition.

3.3 VEGETATION, FOODSTUFFS AND TEXTILES

3.3.1 Compost and Hay

On a cold winter's day, the steam rising from manure and compost heaps is a common sight in the northern latitudes (Figure 3.4). Likewise piles of cotton, hay, silage and manure often steam and occasionally catch fire. It is largely caused by the rapid growth of microbes within the material. Microbial growth results in the production of heat, and if the heat cannot escape, the temperature begins to rise. This source of heating cannot continue indefinitely since most microbes are killed off at temperatures exceeding 70 °C. Therefore it may come as a

Figure 3.4 Compost steaming on a winter's day.

surprise to learn that such piles can often burst into flame. The phenomenon has been known since antiquity. The Roman writers Columella and Pliny the Elder gave advice on the cutting of hay to reduce the possibility of fires. It had been observed that damp hay was particularly at risk.

Within a compost pile temperatures often attain 60 °C and occasionally reach 80 °C or more as a result of microbial activity. Composts and manures normally contain adequate amounts of water and nutrients to ensure rapid microbial growth. Microbes are remarkable for their ability to exploit different forms of organic matter. Growth can occur under a wide range of conditions. Almost any type of organic material can provide an energy source, but as far as compost piles are concerned, grass clippings and other trimmings seem particularly effective.[3] If free oxygen is present, microbes carry out aerobic respiration in order to secure their activities. Oxygen provides a particularly efficient way of converting organic matter into chemical and other forms of energy. When free oxygen is absent, some microbes are able to respire anaerobically using combined forms of oxygen. In this way they successfully exploit most kinds of organic matter that we know about. Microbial respiration is far from 100% efficient in oxidising organic matter to produce chemical energy for growth and maintenance. Most of the remaining energy is converted into heat. As noted above, temperatures >80 °C kill most forms of life, but some recently discovered microbes, known as extremophiles thrive at temperatures beyond that of boiling water. Since microbes are small they are efficiently dispersed and such forms can probably be found in all compost piles.

The heating afforded by respiration was first clearly demonstrated in 1830 by the German chemist Heinrich Goeppert.[4] He noticed that grain allowed to sprout in a wooden box could reach a temperature of 50 °C. This was due to rapid respiration during germination and not to microbial activity. However, in hay and compost piles we may assume that the only growth that is occurring is microbial. Good evidence for this has been provided by a series of experiments on sterilised- and non-sterilised hay, which when wetted only heated up in the non-sterilised material. A temperature of 80 °C is a long way off the temperature of a burning pile, so how could combustion occur? The question has been asked many times over the years and has generated much discussion and controversy. One explanation is that above 80 °C, additional non-biological exothermic reactions take over. This could increase the temperature up to the ignition point. Many volatile plant and microbial products would be present in the pile which could react catalytically with the hot solids resulting in rapid oxidation. Some of these volatiles are known to be flammable. Methane is an example although this gas has a high auto-ignition point. John Scott Haldane and R. H. Makgill[5] measured the uptake of oxygen and the production of carbon dioxide in wetted hay at a range of temperatures. These researchers provided convincing evidence for microbial heating in hay and they continued to observe oxygen uptake at 90 °C. At this temperature the majority of microbes should have expired, suggesting that non-biological oxidation was then in progress. With such a hot compost pile, most if not all of the oxygen at the centre of the pile would soon be exhausted but it would still be present at the margins. Here rapid reactions leading to fire might occur. Most reports of combustion are from large compost piles within which the heat is more readily retained. This is due to the poor conductivity of the material. Several estimates have been made of the size necessary to initiate spontaneous combustion. Models have been published based upon the activation energies for combustion of composted material. One of these suggests that compost piles need to reach a volume of about a cubic metre in order to heat up significantly while for actual combustion the piles would need to be several times larger. A recent report from the UK however, which described the complete destruction of a house by fire, showed that combustion started in a small compost bin that was rather less than one cubic metre in volume. The fact that the fire occurred in mid-summer and the compost was covered in black plastic may have had some bearing on the incident.

Another important factor leading to these fires is the compost moisture content. There appears to be a critical range of between 20 and 45% moisture by weight within which these fires are initiated. Above that level it has been assumed that the amount of contained water is sufficient to cool the pile through evaporation. A high water content may also result in a higher water vapour pressure within the pile, competing or inhibiting the production of organic volatiles. Below a 20% water content it is assumed that microbial growth is inhibited leading to reduced temperatures.

The spontaneous combustion of hay ricks as opposed to compost most likely results from similar causes. In Switzerland it has been reported that about 0.5% of hay can be lost in this manner. In North America, losses were reported to run into millions of dollars in the 1920s. Piles of hay differ little from compost heaps made from grass cuttings. They are often harvested under similar conditions but the strands of hay are typically longer, probably allowing a greater ingress of air. The other major difference is economic – large sums of money can be lost if a hayrick ignites as this often destroys the building containing it. Serious interest in the phenomenon began in the 18th century when the Dutch scientist Herman Boerhave attempted to analyse the material remaining after a spontaneous hay fire. He obtained a '*water impregnated with an urinous spirit*' and managed to identify ammonia and ammonium carbonate in the liquor. Neither of these substances is readily flammable however. The earliest detailed account of hay combustion was given by the German chemist H. Ranke[6] in 1873. He observed a smoking pile of stored hay and proceeded to excavate it to find the cause. The mound was 7 m high and cool at the surface but showing evidence of water condensation. At a depth of 1.5 m it was deep brown in colour, strongly odorous and hot. Sparks were also noticed at this depth. A careful excavation resulted in a diagram of the state of the pile (Figure 3.5). Ranke then undertook some experiments. He heated some hay to 250–300 °C when it carbonised (blackened). Providing the material was sealed as it was heated, to contain the volatiles, it became red-hot in places on exposure to the air. But if the volatiles were allowed to escape before exposure, this did not occur. It was therefore assumed that the combination of volatiles and the carbonised remains of the hay was necessary for spontaneous combustion. Browne[7] suggested that some form of catalytic action may be responsible for the combustion.

A similar process may be responsible for both hay and compost fires. It has been suggested that cellulose, a major plant component, is initially carbonised *via* microbial or non-biological heating. The resulting carbon acts as an adsorbent for oxygen

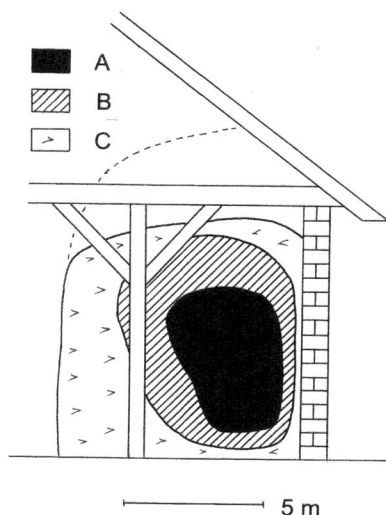

Figure 3.5 Vertical section through a hay stack that has undergone self-ignition. (A) completely charred, (B) partly charred, (C) undamaged. Broken line shows the original profile of the stack. (Adapted from ref. 7).

present in the air, permitting the oxidation of a volatile fuel. This then raises the temperature sufficiently to start a fire. While the hypothesis is plausible there remain a number of uncertainties. In particular, a temperature of 80–90 °C may not be sufficient to carbonise cellulose in the time available. An alternative suggestion is based upon the known pyrophoric properties of finely divided carbon but it is unclear whether carbonisation would be capable of producing carbon in such a state. Plant chloroplasts contain linolenic acid, an unsaturated and reactive oil known to be involved in spontaneous combustion (see below). Whether sufficient oil exists after microbial degradation however is debatable.

A surprising discovery was made by Laupper[8] who in 1924 concluded that pyrophoric iron was responsible for the ignition. He found that iron was associated with carbonised hay. This is to be expected since this element is present in all living cells. But when the iron was removed with acid, the material was no longer capable of igniting. Replacing the iron as rust, however, resulted in ignition, providing the material was heated to 280 °C. After many studies he prepared a detailed description of the changes that he believed took place in hay as the temperature was raised from 35–340 °C. He noted a number of gaseous substances were produced including hydrogen sulphide, formic acid and furfural, all of which are flammable. He also noted ammonium nitrate, a well-known explosive, but its role if any in hay combustion remains unclear. There have also been reports of fragments of discarded iron being found in charred cavities within piles lending support to Laupper's observations. Other research found that ammonia may be indirectly linked to combustion events. It was discovered that fires tended to be more common in hay containing clover or alfalfa, crops that have a high nitrogen content. Decomposition leads to ammonia formation. This will dissolve in water to produce an alkaline solution. It was already known that sugars react with alkalis to produce reactive molecules. They can be rapidly oxidised by the air (see linseed oil, below), and it is possible that combustion in hay could result from this. It is also known that combustion is much less common in straw. These culms of harvested wheat contain little nitrogenous material. Damp straw however has found an unlikely use that suggests that it too may be involved. The chance discovery that straw bales left in standing water can kill toxic cyanobacteria led to the finding that hydrogen peroxide is formed during its degradation. It appears that peroxide, a powerful oxidising agent, toxic to algae, can be formed by the interaction of wet straw, sunlight and humic matter during its decay. However the quantity produced is small and the straw would normally be too wet to take fire.

Manure also catches fire. An account from the State of Virginia in the United States describes that *On a hot September day a fire was noticed in a large pile of horse manure. The mass was packed so firmly that progress was slow. In one of the hottest sections a forkful of material was removed and spread out on the ground. The straw was hot and steaming. About one minute later the steam had changed to smoke, which increased in density until after about 3 minutes the material glowed a fiery red.* A maximum temperature of 167 °C was recorded close to the pile surface. The phenomenon shows much in common with hay fires although combustion appears to have occurred close to the surface in this case.

3.3.2 Foodstuffs – Flour and Sugar Dusts

Ground flour, consisting mainly of starch, is combustible. There have been many reports of flour explosions, when the fine dust mixed with air is ignited. One of the worst of these occurred at the Tradeston Flour Mills in Glasgow in 1872. Grain was being continuously fed to a pair of millstones for grinding into flour when it appears that the grain flow stopped. The wheels kept turning and eventually they met and sparks were created. These gave rise to a small explosion in the vicinity. It appears that this initial explosion shook the building and allowed clouds of flour dust to fill it. A huge secondary explosion ensued leading to the complete destruction of the building and great loss of life. Such explosions are often thought to result from machinery. Electric sparks up to 1 m long have been observed for example in grain elevators. Dust has been defined as a state of matter where the particle size is less than half a millimetre. Such materials are easily suspended in the air, especially where their density is low and their dimensions are much less than half a millimetre. These explosions occur in all kinds of flammable dusts and have been known from sugar, corn starch, paper, wood and polyethylene.

Sugar refining involves the clarification of raw cane- or beet sugar to remove impurities and colour. Once this has been achieved the sugar solution is evaporated and crystallised. Finally it is gently ground and packaged. It is in this last stage that explosions have happened. In 2003 there was an explosion at the British Sugar plant at Cantley in Norfolk.[9] It occurred in an elevator tower when welding was taking place. Fortunately the explosion was not devastating allowing the course of events to be followed. The plant was idle during the incident. The welding was soon suspected and sections through the weld were taken. In addition charred sugar was found in the inside of the tower. In fact much of the tower was encrusted with a thin layer of sugar. This provided the ignition source for the explosion. Since there was no activity at the factory it was assumed that workmen had disturbed the tank leading to a cloud of sugar dust which then ignited explosively.

A devastating explosion occurred in the Imperial Sugar Refinery at Port Wentworth, Georgia in 2008. In this case it appears that sugar fines built up along an enclosed conveyor system and came into contact with an overheated bearing. An explosion took place in the confined space leading to further, more devastating explosions in the storage facilities above.

The auto-ignition point of cane sugar is low at about 170 °C. During its decomposition, a flammable gas, hydroxymethylfurfural, is also given off. Several causes for these explosions have been suggested. They include machine malfunctions, friction, electrical faults and spontaneous combustion. A German study found that over half of reported dust explosion incidents were down to these causes.

Another dusty material is Lycopodium powder. This consists of the dried spores of lycopods, plants related to ferns. The dust, although not known to self-ignite, is highly combustible and for many years was 'stock in trade' for magicians who would blow some of the powder into a candle causing a bright flash and small explosion.

3.3.3 Textiles

The fine cellulose fibres of the cotton plant should be a good contender for spontaneous combustion. In fact numerous reports of this have come to light owing to its widespread use in the textiles industry.[10] The cotton plant is a shrub whose seeds develop long hairs in such abundance that they may be harvested for textile manufacture. The hairs are a dispersal mechanism. They are caught by the wind allowing the seeds to be carried long distances. The hairy seed masses, known as bolls, once harvested, are placed in a cotton gin that separates the fibres from the seed. Cotton is then transported in large bales that measure approximately $1 \times 0.5 \times 0.5$ metres. In the cotton industry there have been reports of bales 'fire-packing' over the years. Such bales appear normal on the surface, but when broken apart, weeks or even months after packing, they reveal a smouldering mass of cotton within. Bales have been held responsible for a number of serious fires in North America. The phenomenon prompted experiments to determine the conditions leading to fire-packing. One method involves a 'self-heating' test, which has been used for a wide range of potentially flammable materials. A small sample is placed in a wire cage and introduced into an oven held at 140 °C for 24 hours. The temperature of the sample is monitored and the material is classified as spontaneously combustible if it either self-ignites or its temperature continues to rise above 200 °C. Cotton was found not to be spontaneously combustible under these conditions. Further calculations based upon the amount of air contained within the packed bales appeared to confirm this. In order to sustain a fire in a bale, a source of oxygen is required and initially this will be confined to the bale itself. Bales contain about 70% air and within an average bale, this is only sufficient to consume about 100 g of cotton. This is a small proportion of the whole. Based upon this evidence cotton does not appear to be particularly at risk of self-ignition. There is further evidence to support this. First, baled cotton has a low water content and this would inhibit any microbial growth. Second, any microbes within the bales would be starved of nutrients. Cotton consists of practically pure cellulose, but microbes need significant amounts of other essential elements, particularly nitrogen and phosphorus. In their absence they would fail to grow unless the cotton was contaminated with other matter. Fire-packing however demonstrates that bales can burn from within. In former times, the cotton bales were packed less densely allowing more opportunity for their oxidation. It also appears that most, if not all of these fires were the result of sparks and other hot fragments of material. The sparks were believed to be caused by static electricity from the cotton gins. The careless disposal of lighted cigarettes was also reported. Care also has to be exercised in retaining bale purity in case non-biologically-mediated oxidants contaminate the product. Vegetable oils for example can promote spontaneous ignition of cotton below 200 °C.

The combination of natural oils with cotton can be lethal. The oils responsible belong to a group known as the omega fatty acids. These consist of long hydrocarbon chains ending in a weakly acid group. They contain one or more double bonds between the carbon atoms of the chain. These double bonds make the oils reactive and this is one of the reasons why they are useful. For example, linoleic acid, obtained from flax is used in varnishes. It forms a firm dry film on a painted surface. Upon drying, the acid is converted by the oxygen in the air to produce a strong cross-linked film, as the gas attacks the double bonds. This reaction and other more complex ones are responsible for the evolution of heat and may cause fires under certain conditions.[11] Such fires were common in the 19th century when the oiling of textiles was practised to assist in waterproofing the material.

Fires were also reported by painters and decorators in the United States who used cotton waste to wipe up linseed oil spills. This resulted in the employment of the Mackey Test first documented in 1895. Studies showed that the mass of material was an important factor for self-ignition. Further tests using different quantities of materials led to the Frank-Kamenetski evaluation in the 1930s. Linseed oils are often refined or mixed with other materials so their properties vary a good deal. In fact all unsaturated oils are a fire risk and these include the fatty acids found in olive oil and fish oils.

Adding oil to cloth both increases the surface area of the oil and its exposure to the atmosphere. The cloth itself plays no part in the heating but of course it is also combustible – the auto-ignition temperature of cotton cloth has been observed as 276 °C. Iron oxides are said to hasten the oxidation and at least some water appears to be necessary. A single handkerchief-sized square of cloth soaked in one type of linseed oil caught fire within 6–8 h at ambient temperature. Even a few grams of oil suitably placed can lead to combustion.

Linseed oil contains linolenic acid, an unsaturated compound containing several double bonds as indicated in Box 3.1.

A remarkable incident involving cotton transport was that occurring on the sailing ship the Earl of Eldon. This full-rigged ship made a fateful journey from India to England in 1834. A stirring account published in Wetherill's *Ancient Port of Whitby and its Shipping*[12] was given by one of the passengers, Mr. T. T. Ashton of the Madras Artillery.

On the 24th August, 1834 I embarked on board the ship Earl of Eldon of London, 600 tons, Captain Theaker, at Bombay. She was the finest and strongest ship of the trade, and was cotton loaded. On September 26 we got onto lat 9°27' south and between 70° and 80° east and began to anticipate our arrival at the Cape [of Good Hope]. On the morning of the 27th I arose early and went on deck finding one of my fellow passengers there. We perceived steam rising from the fore hatchway. I went

---

BOX 3.1   LINOLENIC ACID AND ITS DOUBLE BONDS

Some hydrocarbons are described as unsaturated. This means that the molecules can accept additional hydrogen atoms. For example, the simple compound ethene has the molecular formula C_2H_4. It can accept a further two hydrogens to form ethane, C_2H_6. This is because ethene has a carbon–carbon double bond. Its structure is written as $H_2C\!=\!CH_2$. When this bond is broken, two more hydrogen atoms can attach. Linolenic acid is not a hydrocarbon because it contains some oxygen atoms but it does have three double bonds as shown below.

The structure of linolenic acid:

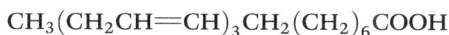

$$CH_3(CH_2CH\!=\!CH)_3CH_2(CH_2)_6COOH$$

The formula below shows that linolenic acid contains three sets of double bonds, $CH\!=\!CH$

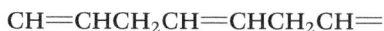

$$CH\!=\!CHCH_2CH\!=\!CHCH_2CH\!=\!$$

Carbon–carbon double bonds are more reactive than single carbon bonds. This is because the atoms share more electrons meaning that the region surrounding the two carbons becomes 'electron rich'. Oxidising agents seek out such electrons. Agents such as oxygen can therefore attack them and remove some of the electrons. In the case of linolenic acid significant heat is evolved in the reaction and this can lead to ignition. In cases where the oxidation rate is slowed down, caused by reduced access of the oxidant, bond-breaking results in the formation of a tough cross-linked polymer with a complex structure. This feature is made use of drying oils in artist's oil paints and window putty.

---

down to dress and about half past six the captain told me that the cotton was on fire. At eight o'clock the smoke became thicker, and before nine we discovered that part of the deck had caught fire. The captain ordered the boats to be got out and stocked in case of necessity. [By] three o'clock we all got onto the boats, the captain being the last just as the flames were bursting through the quarter deck. When we were about a mile from the ship she was in one blaze, and her masts began to fall in. The sight was grand, though awful. Between eight and nine o'clock her masts had fallen and she had burned to the water's edge. Suddenly there was a flash and a dull heavy explosion. Her powder had caught. A few seconds and all was dark, and the waters closed over her.

Sad was the prospect before us. There were in the longboat, 23 ft by 7 ft, 25 persons and in each of the other two boats ten individuals. We were by rough calculation about 1000 miles from Rodrigue[s][Island]. About 11 o'clock, having humbly committed ourselves to Providence, we rigged the boats and got under sail.

After many privations, the boats eventually made landfall without loss of life at Rodrigues Island in the Indian Ocean, 14 days after boarding the longboats. Spontaneous combustion has been cited as the likely cause of this fire but doubts remain. However the cargo was apparently damp, and if the bales contained other materials self-ignition would certainly be likely, particularly for a large amount of material confined in a ship's hold in a tropical climate.

There are also reports of laundered clothes catching fire after having been removed from tumble dryers. Ignition has been reported at around 90 °C. In this case there may be a direct oxidation of the laundry, but it could also be due to traces of contaminants such as drying oils, bleach or urine in the material. Another potential hazard with these dryers is static electricity. The sparks have been known to start fires.

3.4   FOSSIL FUELS – COAL

Vegetation sometimes accumulates in large masses to be buried rapidly under sediments in lakes and the sea. If this is followed by compression and heating for a long period, coal is the result. Most of the hydrogen and oxygen present in the organic matter is lost, but carbon, the other main constituent, is largely retained. The resulting material has a carbon content ranging from about 70–95%. Although the carbon content is high, most of it remains combined with some of the original hydrogen and oxygen to produce a complex, amorphous colloid. Some of the constituents are volatile and easily ignited, making coal a valuable fuel that is easy to burn. It also has a high calorific value, a property that was recognised from early times.

There is evidence of coal burning from prehistory and in the UK where coal is found in abundance, it has been used at least since the early medieval period, eventually fuelling the Industrial Revolution. Coal is classified according to its rank. The rank of coals is determined by several factors, the most important being its carbon content and its constituents. Anthracite is of the highest rank with a carbon content of up to 95%. Lower rank coals are more common and contain a significant proportion of volatiles. Sub-bituminous coals are considered to be most at risk of spontaneous combustion and these are rich in fossilised plant fragments called macerals. They include vitrinite, the remains of woody plants and liptinite, consisting of spores, waxes and pollen.

3.4.1 The Yarmouk Enigma

In parts of the remote Yarmouk Valley in the northern Maquarin District of Jordan are some peculiar springs that are highly alkaline and caustic. They are formed by the passage of water through a rare mineral, portlandite. This mineral is no more than the familiar slaked lime, one of the ingredients of Portland cement. Slaked lime is still made on a large scale by heating limestone (calcium carbonate) to a high temperature, driving off carbon dioxide. The result is calcium oxide (quicklime). When lumps of quicklime are placed in water there is a violent reaction with the formation of slaked lime (calcium hydroxide) as noted in the previous chapter.

In the Yarmouk Valley, portlandite is found associated with bituminous limestones that have an organic matter content approaching 20%. Pyrite is also present. The only reasonable explanation for the occurrence of portlandite in this locality is the spontaneous ignition of the bituminous material.[13] A fire could lead to the baking of the associated limestone to form quicklime. Portlandite would then be formed through the action of groundwater. A good supply of oxygen would be required for the combustion of the coal, possibly through the development of fractures following earthquakes or landslides. Perhaps pyrite oxidation also played a part. Detailed fieldwork suggests this took place within the last 600 000 years.

Portlandite has also been found associated with coal in the Chelyabinsk Basin of Russia, and in the bituminous deposits of the Hatrurium Formation of the Negev Desert, Israel, lending support to this idea.

3.4.2 Coal Self-ignition

The spontaneous combustion of coal is a well-recognised hazard of coal mining and storage that must have been observed soon after extraction began and possibly even before. Evidence has been obtained of coal fires dating to the Pliocene period in the Powder River Basin of Wyoming, so named from the evolution of sulphur dioxide gas from the deposits. There are numerous references to coal and lignite combustion in Italy. One of the earliest is that of Appolonius Rhodius who mentioned that Argonaut sailors *'entered deep into the stream of Eridanus; where once smitten on the breast by the blazing bolt, Phaethon half-consumed, fell from the chariot of Helios into the opening of that deep lake; and even now it belcheth up heavy steam clouds'*. The site is located in a depression near the town of Spina but there are no recent records of fires in that region. The first coal was probably exploited by quarrying where the seams were close to the surface. At another Italian site there is an early reference to the taste of wine being affected by the burning soils. With close proximity to the open atmosphere, these deposits would be readily permeated by air and are known to be particularly prone to burning.

In the 1600s deposits of alum shale at Salzweiler, Germany, were once 'burned' *in situ* to yield their aluminium sulphate by igniting the underlying brown coal. The coal was still recorded burning 200 years later.

Mount Wingen in New South Wales, Australia, contains perhaps the earliest known example of spontaneously combusting coal. Here a seam of coal close to the surface has been smouldering for perhaps as long as 6000 years. It continues to move south at the rate of about one metre per year. It was known to the aborigines and the early British settlers regarded it as a smoking volcano.

Near Dunrobin Castle on the east coast of Scotland, Thomas Pennant in 1772 observed[14] '*small strata of coal three feet thick dipping to the east, and found at a depth of about 14 to 24 yards. Sometimes it takes fire on the bank, which has given it an ill name, that people are very fearful of taking it aboard ships. I am surprised that they will not run the risk, considering the miraculous quality it possesses of driving away rats wherever it is used*'. Such occurrences must have been commonplace once mining began on an industrial scale. In the Birmingham area where coal seams were often close to the surface, spontaneous combustion was once well known with the associated gas igniting at the surface. The Black Country village of Moxley, Wolverhampton, was once the location of the 'Fiery Holes'. The area was formerly a coal mine. It was sunk to access the famous 'Ten Yard Seam'. The coal however had a high sulphur content and incidents of spontaneous combustion were reported to be of frequent occurrence. The Fiery Holes public house nearby remains a testament to these occurrences.

Coal mining underground usually followed the pillar and stall method prior to the use of modern coal cutting machines. The coal seam would be removed in such a way as to retain substantial pillars of coal to support the roof. This was an inefficient but necessary method to reduce roof falls and meant that only about one third of the seam was extracted with the remainder in the pillars. Even the removal of this amount of coal leads to a significant increase in stress in the remaining coal. There are reports of frequent ignition events originating in pillars, known as 'gob fires'. Although there is no direct evidence, it is likely that these fires began as a result of the increase in stress caused by coal extraction. The stress produces a local movement and crushing of coal, with subsequent friction bringing the material to the ignition point. The presence of pyrite and efficient ventilation may well exacerbate the problem.

Within coal mines, the flow of air can be a significant factor in determining the severity of any combustion event. High flow rates can dissipate the heat reducing the risk while low rates tend to increase it. Fires of this nature are far from rare. Sudhish Banerjee made a comprehensive study of the phenomenon in some Indian coal mines.[15] He found records of 158 underground and 212 surface fires up to the 1970s. Similar numbers have been recorded from Italy and it is likely that all significant coal mining areas are affected. However, not all coalmine fires have been correctly identified as a result of spontaneous combustion. The Blackburn Standard of March 5th 1851 reported smoke issuing from the workings of Dogshaw Mine near Darwen over several days. Soon a search was undertaken. In a remote part of the old workings an illicit still was discovered. The fire had been lit too close to one of the pillars and had ignited the adjacent seam.

More serious fires have occurred in other mines. A series of spontaneous fires were reported from the William Pit of White-haven, England, beginning in 1911. It led to part of the workings being sealed off to prevent its spread. A series of strong brick seals were built but as the years progressed the fires expanded toward the pit head. By 1938 the situation was getting serious with 20 miners employed to contain the fires. Smoke was entering the active mine through small fractures and holes in the pillars. In 1941 it was decided to drench some rubble in the active workings that had been heated by the fires. The drenching water ran into an old roadway but then disappeared. A few hours later there followed a huge explosion killing twelve miners and injuring a further eleven. This part of the mine had not been affected by firedamp (methane gas), which is frequently the cause of coal mine explosions. At the enquiry it was concluded that the water entered the burning zone where it turned to vapour and reacted with the hot coal to form 'water gas' a flammable mixture of hydrogen and carbon monoxide. This was subsequently ignited on contact with air. The entire area had to be sealed off and abandoned.

Considering the great economic loss, and formerly frequent loss of life associated with the self-ignition of coal it comes as little surprise that the phenomenon has attracted much attention. Work to discover the cause of spontaneous combustion began in the first half of the 20th century. It resulted in the proposal of five 'theories': pyrite oxidation; bacterial action; tectonic heating; water sorption and direct oxidation. Iron pyrites is common in coal and occurs as small crystals or seams with a brilliant silver–gold lustre. In the presence of water and air the mineral oxidises readily with the liberation of heat. The result is a solution of sulphuric acid and iron oxides. Since coal is a poor conductor of heat, this transformation could be expected to result in a considerable rise in temperature. Studies indicate that this indeed happens but is only significant when the pyrites occur as extremely small crystals and at a high concentration. These conditions are uncommon in coal, but may be found in pyrite mines, where there have been occasional reports of fires resulting from their spontaneous ignition. A catalytic role for iron has also been suggested as in the case of compost piles. Iron pyrites is an iron(II) sulphide while iron(III) sulphide Fe_2S_3 is pyrophoric but does not appear to occur in nature.

Microbial activity has already been seen to raise the temperature of compost piles. This prompted studies using coal. Bacterial cultures have been inoculated into finely powdered coals and the temperature increase was measured. Bacteria appear capable of utilizing some of the substances present, but the temperature rise so far observed has not been considered sufficient to initiate combustion.

Coal seams can be disturbed by earth movements leading to the crushing and movement of the particles. There is plenty of evidence in coal mines showing disturbance on a large scale but most of these changes occurred in the distant past. However mining activities often trigger renewed movements leading to localised disturbances. They may heat coal as well as crush it, releasing dust. Rapid collapses or crushing leads to adiabatic heating that could easily exceed the ignition temperature although the high temperatures obtained are usually of short duration. Crushing also exposes a larger area of coal to the atmosphere, potentially increasing oxidation. It was suggested that the heavy overburden in coal pillars produces small cracks allowing oxygen ingress followed by ignition. It may be this process that is responsible for the lights occasionally reported from abandoned coal mines. A further hazard present in deep mines but not affecting coal stored at the surface is geothermal heating. Mine temperatures are invariably higher than the temperature at the surface, particularly when they are deep underground.

The fact that fine coal particles are oxidised much more rapidly than large lumps has been confirmed by laboratory studies. The air was admitted into a tube containing a coal sample and then extracted and analysed for changes in its composition. In parts of Italy, peat deposits are known to self-combust. Some were found to overlie natural gas seeps which are likely to exacerbate the problem.

Coal displays a complex chemical relationship with water.[16] All coals contain water, some of which is in the free state occupying small crevices and voids, while the remainder is physically or chemically attached to the coal surface. Since coals have a high surface area even when they are uncrushed, a large amount of water may be bound up in this way. Upon admission of water into coal, a considerable rise in temperature is often observed. This is known as the 'heat of wetting'. It is probably caused by an exothermic reaction at the coal surface. If water is admitted as vapour, even greater temperature rises occur resulting from its condensation onto surfaces.

All of the above processes probably contribute towards the self-heating of coal but the relative significance of most of them is unclear. However there is plenty of circumstantial evidence implicating moisture and crushing, since coal fines possess much greater combustion activity than large lumps.

A century ago, John Haldane found that oxygen admitted to coal exists in at least two forms – as a solution in water and in a chemical combination.[17] Haldane was a Scottish medic who had a strong interest in coal mine explosions and their cause. He was the first person to identify carbon monoxide as the lethal component of 'afterdamp' produced after the explosions.

More recent work has indicated the presence of an unstable intermediate substance that has been termed a 'peroxy complex' as its formation rate is similar to the change in 'peroxide number' of certain foodstuffs and oils. The breakdown of the complex in the presence of water is thought to release free radicals, highly reactive molecules that can increase the rate of oxidation manyfold. Interestingly, the amount of peroxy complex in coals depends upon their rank. It is the lower rank coals that possess the highest peroxy values and it is these that are most liable to self-ignite. The role of water here may also be crucial. Many fires start when wet coal begins to dry out, gradually exposing the large internal surface and its peroxy compounds to the oxygen of the air. At the same time the thermal conductivity of the coal is reduced allowing greater heat retention.

One technique that has thrown much light on the process of coal oxidation involves measuring the loss in mass and production of gases as the temperature is increased. Small samples of coal are placed inside tubes and subject to an air flow at temperatures ranging from ambient to several hundred degrees Celsius. Changes in composition for a wide range of coals can then be investigated. A variation of this theme is called differential thermal analysis or DTA. Here self-heating can be measured with great accuracy over a range of temperatures. Studies have shown that self-heating begins at 100–150 °C for some coals while at lower temperatures, cooling results, probably as a result of water evaporation. Almost identical changes can be detected in the loss of mass and the production of oxidation products (the gases carbon monoxide and carbon dioxide). These and similar tests are conducted on coals to predict their liability to self-heat.

As coal begins to self-heat underground, a range of volatiles are driven off and produce 'fire stink' giving the miners an early indication of a problem. Several detectors have been invented. One of the earliest of these were the 'stench agents', small vials of intensely odourous compounds such as mercaptans that vaporise on warming. These were placed in different areas of the mine and would give warning to the miners that changes were afoot. They were superseded by a range of more specific detectors for carbon monoxide and later by thermal infra-red detectors. The latter are particularly useful in detecting the source and intensity of the heating. There is also one other way in which coal might self-ignite in a mine. Some coals contain the element germanium, and germanium hydride is spontaneously combustible. This hydride occurs as a gas, and the reducing conditions found in coal seams may be just sufficient to cause its formation.

If a fire manages to take hold in a mine, the consequences can be felt for generations to come. To save the mine, attempts are made to block off all sources of oxygen so that the fire is starved. It has been found that 'flaming combustion' ceases below about a 12% oxygen content in the air but a slow oxidation can continue at much lower levels than this, at about 1–2% oxygen. It is not an easy matter to starve a mine fire entirely of oxygen since the ingress points are not always obvious. Underground fires may continue for years. A fire that started in 1862 at the Astor Mine of Allegheny County, Virginia, was not extinguished until the early 20th century. The bustling town of Centralia, Pennsylvania was reduced to a ghost-town after large fires developed in the mines below. They were first thought to have started spontaneously, but it is now considered more likely that they were caused by the transfer of heat from a surface dump that caught fire. The fires have raged underground since 1962. But there is little evidence of their existence on the surface. In 1979 a local petrol station owner went to check the fuel levels in the underground tanks and was shocked to discover that the temperature had risen to nearly 77 °C. A few years later a young boy was saved from falling around 40 m into a smoking sinkhole. The town was subsequently abandoned to the fire.

Piles of extracted coal are prone to self-ignition and the process is easier to examine than in mines. Stacks more than 2 m high are usually affected. It has been found that the build-up of heat is greatest 1–2 m inside the stack. The high porosity of the material permits the upward escape of volatiles resulting in a chimney-effect, forcing air into the bottom of the stack accelerating the oxidation (Figure 3.6). Fires in coal wagons are also encouraged by the additional supply of oxygen caused by the moving train. The Titanic had huge coal-powered engines and the holds contained around 100 tonnes of coal each. It is known that a spontaneous fire occurred in one of these on the ship's maiden voyage. The resulting damage to the steel is thought by some to have made its hull more prone to rupture in the subsequent collision with an iceberg.

Even the spoil from coal mines may contain sufficient organic matter to combust. In former times, children used to play on the huge waste tips surrounding the mining town of Wigan in Lancashire. Here they took advantage of the warmth provided by

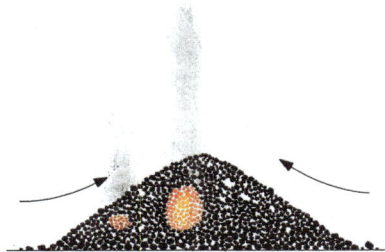

Figure 3.6 Illustration of spontaneous combustion in a coal pile *ca.* 2 m high. Arrows show direction of air flow resulting from convection.

combustion beneath the surface during the cold winter months although it meant suffering the sulphurous fumes that were often emitted. Although such exposure is harmful to human health, the fumes were once recommended as a cure for Whooping Cough. Two large spoil heaps covering 49 ha were stripped in the 1980s for landscaping when hot-spots were discovered, melting the tyres of the excavators. Further examination revealed temperatures of 600–900 °C due to coal waste oxidation. This could only be prevented by sealing the tip with fly-ash and a covering of gas-impermeable compacted clay. In the spoil tips of the Shirebrook Colliery in Derbyshire temperatures of more than 300 °C were recently found. This tip had probably been burning for about 50 years. The extent of the heating was revealed by thermal imaging from an aeroplane.

Wooden ships are particularly prone to fire. On long journeys, their holds were often stuffed full of goods, many of which would be flammable. Felix Riesenberg, in his account of the ships rounding Cape Horn, drew attention to the *Patmos I* whose cargo of coal began burning in the hold. Despite the icy conditions, the fire soon spread, burning out the oakum between the planks and then melting the pitch. Flames soon spread up through the deck and the fresh water tank steamed as the sailors wrestled with the sails to bring her closer to land. Seawater began to pour in, lowering the ship in the water but at the same time helping to quench the flames. Despite their efforts the boat was slowly sinking. They were forced to launch their rowing boat and spent a full day at the oars before Tierra del Fuego came into view. They were later saved by a passing ship.

Coal mine explosions are an ever-present danger to the miner and are usually attributed to build-ups of methane, although coal dust can also represent a hazard. Methane is thought to be sorbed onto coal components and was either formed long ago during the early stages of burial or produced thermogenically. By reducing pressure on the coal, gas will escape. Coal and charcoal dust has also been known to self-heat and ignite. The auto-ignition temperature of methane-air mixtures is high (580 °C) and explosions are most likely to be caused by sparks or naked flames.

Coal dust has occasionally been cited in mine explosions. The Benxihu mine in Liaoning, China, regarded as the worst coal mining disaster is thought to have resulted from coal dust. It was probably exacerbated by poor ventilation but details are wanting. A disaster at Mount Mulligan Mine in Queensland, Australia is better documented and the likely cause was coal dust ignited by an explosive charge. The auto-ignition temperature of coal is in the region 400–460 °C, although temperatures as low as 230 °C have occasionally been reported.

Peat, like coal, has been used as fuel for millennia. Occurring at or near the earth's surface it is easy to extract but is often too wet to use directly. Instead it is stacked in open piles to dry. Although peat has a much lower carbon content than coal, it is readily oxidised. As peat is dried a large surface area is exposed to the air. The Codogoro was once the largest peat bog in Italy but was dug out in the early 20th century for fuel.[18] The peat was placed in piles about 2.5 m high to dry. Piles were separated by ditches to guard against accidental fire but their rapid oxidation was sufficient to scorch the clogs of the oxen used to work the ground. A ranger, who had worked on the English Pennine fells for many years told me that one day, while sitting on a south-facing peat bank in the warm sunshine, smoke began rising close by, to be followed by flames as the peat caught alight. Most peat fires in this area are caused by discarded cigarettes and barbecues and the local fire service did not believe his story. On the Okovango Delta of Botswana, buried peat caught fire when it was exposed by the lowering of the water table during an exceptionally dry period. It left a siliceous deposit in its place.

3.5   FOSSIL FUELS – NATURAL GAS AND OIL

Of all the flammable gases, methane or 'marsh gas' is the best known. Its identity was discovered by Alessandro Volta in the 18th century and has been used as a source of heat and light for centuries. Methane is one of the simplest hydrocarbons consisting of one carbon atom linked to four hydrogens. It is also a light gas such that a balloon filled with it will displace air and rise to a good height. The small size of the molecule allows the gas to diffuse rapidly through water and porous rock. This feature enables it to easily escape into the atmosphere from an underground source. The gas is produced *via* two main processes. The first is the metabolic activity of methanogenic bacteria, and known as biogenic methane. The second is the thermal breakdown of kerogen, a complex mixture of hydrocarbons associated with oil shales, termed thermogenic methane. Biogenic methane may be produced wherever molecular oxygen is absent and organic matter is present, providing the temperature is below about 100 °C. Favourable sites are stagnant organic-rich waters such as those containing decaying vegetation, including the bottom of the sea. Further below ground, where oxygen levels are generally low, methanogenic bacteria may abound where there is a supply of organic matter in the form of fossilised plant and animal remains. The remains can be disseminated among ancient sediments forming black shales, oil shales and even coal. Providing sufficient water and nutrients are present, methanogenesis can proceed apace. Some methanogenesis also occurs on the earth's surface by way of ruminating animals. Two types of methane production are recognised. The first is described as 'acetoclastic' where acetic acid is cleaved to produce a molecule each of methane and carbon dioxide. The second, 'hydrogenotrophic' methane can be produced by microbes when hydrogen reacts with carbon dioxide to produce methane and water.

Thermogenic methane is produced in regions where the temperature lies in the region of 150–250 °C. This is sufficient to cause decomposition of larger organic molecules into smaller ones such as methane, which is thermally stable at these temperatures. Such high temperatures are normally only experienced at depths of 1 km or more. Microbes however are ultimately responsible for most of the methane at or near the earth's surface.

Recently a third potentially significant source of methane has been discovered. Termed abiogenic methane, it is produced by the hydrolysis and oxidation of the mineral olivine, a magnesium iron silicate, by water followed by a reaction with carbon dioxide. Olivine is a major component of the earth's mantle. It reacts with water to produce serpentine with the evolution of a small amount of hydrogen gas. Laboratory experiments have shown that hydrogen is capable of reducing carbon dioxide to methane in the presence of a catalyst. It appears that traces of other minerals in natural serpentines are capable of catalytic action. The reactions take place at moderately high temperatures and pressures and the resulting methane gas, along with unreacted hydrogen may then escape through natural fissures in the rock. Although the process is now well documented it is not possible to distinguish abiogenic methane from the other forms at present so its significance is currently unclear.

Methane levels in the atmosphere, although small have shown a rapid rise in recent years and are of concern as it is a potent greenhouse gas. Bacteria known as methanotrophs break down the gas but the most significant means of removal is in the upper atmosphere. Here reaction with hydroxyl ions leads eventually to conversion into water and carbon dioxide.

The flammability of methane has already been noted. Owing to its abundance, particularly in oil and gas fields, deliberate methane burning is quite common and there are many notable occurrences. Of particular interest in this respect are the so-called 'eternal flames' some of which have been burning for millennia.

There are reports of 'eternal flames' from around the world and they can be categorised into two groups. The first are those deliberately started and maintained, often for religious and cultural purposes, for profit or simply out of curiosity. There are numerous examples, especially in Eurasia. The second is a smaller group that is not deliberately maintained. They are either the result of accidental or deliberate ignition in the recent past or more ancient conflagrations of uncertain origin. These too are often associated with religious or cultural practices.

Examples of the first group include the Zoroastrian Fire Temples in Azerbaijan which were probably built around natural gas exhalations ignited for religious purposes. In most cases the fires are not directly linked to natural emissions. Another example is the Vesta fire in Rome. It was linked to the fortunes of the City and the State. This fire was tended by the Vestal Virgins but was not fuelled by gas and apparently went out quite often. When this happened, it had to be lit ceremonially by focussing the sun's rays using brass mirrors. The Vesta cult continued for many years but the fire was put out for good in 394 CE. Methane emissions in other parts of Italy are however common but apparently not utilised for these purposes.

In the Republic of Ireland the sacred Fire of Brigid of Kildare was once famous throughout the land. It is thought to have originated in pre-Christian times and was only extinguished during the purge of the monasteries by Henry VIII in the 1500s. The practice has recently been revived but natural gases were never likely to have been involved. There are no records of similar fires in the UK. Here the nearest to an 'eternal flame' would be the 'Burning Well' recorded from the Wigan coalfields in Camden's *Britannia* of 1695. It was extinguished many years ago and its precise location remains a mystery. Another fire with a long history is that in the Buddhist temple of Daisho in Japan, thought to have been started in 806 CE. A number of more recent 'eternal flames' have been built into modern structures to celebrate disasters, wars, significant people or events. They are all fuelled artificially using sources of flammable gas and include that at Anfield Stadium, Liverpool, to commemorate the Hillsborough Disaster, the Arc de Triomphe, Paris (World War 1), and that at Raj Ghat, India, for Mahatma Gandhi.

In the second group are a number of remarkable fires. The celebrated Chimaera's Flames situated close to the ancient city of Olympos in the Beydaglari National Park of Turkey must be one of the oldest. These arise from fissures in rocks containing limestones and serpentines on a rocky hillside. They are thought to be the inspiration for the *'Fire which never goes out'* in Homer's Iliad written around the 8th century BCE. If the process of serpentinisation is occurring in the vicinity, the source of the methane may be abiogenic.

Another example with a similar pedigree is Baba Gurgur near Kirkuk in Iraq. It is thought to have been worshipped since 600 BCE and is steeped in legend. The fire is contained within a shallow bowl about 40 m in diameter. It is set in the desert sands with the yellow flames up to about 0.5 m high occupying about 10% of the area. The site is situated in one of the world's largest oil fields. Its origin is unknown and access is currently difficult. In the text of *De Mirabilibus Auscultationibus* (*ca.* CE 117), it is noted that the burning fires of Persia (Iran) were so large that kitchens were constructed close to them. Such fires are known to have existed in the 3rd–5th centuries BCE. Another early and well known example is that of the Oracle at the Temple of Delphi in Greece where flammable gases once rose through the rocks and readily ignited.

Two of the most impressive fires began more recently. The Yanar Dag or 'Burning Mountain' is sited on a peninsula near

Baku in Azerbaijian. It was started accidentally in the 1950s and the ignited gas appears to arise from a narrow fault along a sandstone scarp. The flames run for some distance along the fissure and can reach a height of more than one metre. Earlier examples of natural gas fires are known in this region. The second is Darvaza Crater in Turkmenistan, also known as the 'Door to Hell'. The crater is 69 m across and formed when the area was being explored for gas in 1971. The ground collapsed into a cavern and began emitting methane. Oilfield gas emissions are often lit deliberately to dispel the gas safely but they usually subside after a few weeks or months. In this case the flames continue to make a stunning night-time display around the crater and have become a major tourist attraction (Figure 3.7). The bowl-shaped depression probably aids combustion as it would tend to reduce the loss of heat from the fires through radiation. Other smaller but no less impressive displays in a religio-cultural context occur in other parts of the world. These include the Mrapen of Java, in the village of Manggarmas, where a natural gas flame was lit prior to the 15th century. There is an annual Buddhist ceremony associated with it although the site of the gas emission has been much modified over time. The Jwala Ji shrine at Muktinath, Nepal, is situated in a small cave where methane and water rise together. The burning mixture gives the impression that the water is on fire. A similar phenomenon occurs at the 'Water and Fire Cave' in Guanziling, Taiwan. Here the methane is said to have been burning for around 300 years. Methane is also produced near the hyperalkaline springs at Acquasanta in Italy. These springs are often associated with serpentinisation. At some

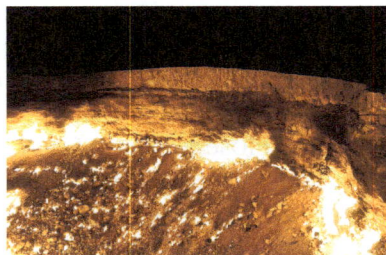

Figure 3.7   Darvaza Crater, Turkmenistan. Seepages of methane appear to coincide with bedding planes in the sediments. Reproduced with permission from John Pavelka.

time before 1400 CE shepherds were reported to have seen lights among rocks in the river, perhaps the result of spontaneous combustion. A Christian sanctuary was later built nearby. Pliny described burning gases near Etna, and the fields of Aricia near Rome have escapes of natural flammable gases. The Roman author Claudius Aelianus mentioned a spring near Apollonia on Sicily that emitted flames.

Eternal flames in the New World appear to be few. Without doubt the best known is that at Eternal Flame Fall in New York State, USA. This is to be found in Chestnut Ridge Park in a small enclave below a waterfall. Although modest in size, and apt to extinguish from time to time, it is a local tourist attraction and unusual in containing significant amounts of the hydrocarbons ethane and propane in addition to methane.

While spontaneous combustion cannot be ruled out for some of these fires, those few not started by man may have come into existence *via* lightning strikes. Anecdotal accounts have occasionally suggested that natural caves and springs attract lightning. This is thought to be due to the enhanced conductivity due to the occurrence of radon and its fission products in the air surrounding them (Chapter 5). The products ionise the air in their vicinity which could make strikes more likely. The famous caver, Norbert Casteret, noted that trees in the vicinity of caves in the Pyrenees were more frequently struck by lightning but hard evidence appears to be lacking.

3.6   OIL SHALES

Oil shales are rich in organic matter. They do not contain oil but kerogen, a dark organic substance that yields oil on distillation. Most sedimentary rocks contain kerogen but only a few places are rich enough to yield economic amounts of fuel. It is probably the most abundant form of organic carbon in the earth's crust. Kerogen, whose name means 'oil or wax-bearing', is a complex mixture of hydrocarbons and polycyclic aromatic compounds. The hydrocarbons are mainly long-chain paraffins, carboxylic acids and cyclic hydrocarbons known as naphthenes. Nitrogen, oxygen and sulphur are also constituents, the latter sometimes forming cross-linked compounds giving a rubbery texture to the material.

3.6.1 Smoking Hills

One of the earliest known examples of oil shale fires must be that at Smoking Hills of Franklin Bay in the Northwest Territories of Canada. Archaeological studies indicate that the site has been burning for millennia and perhaps as far back as the end of the last glaciation. The area consists of low cliffs of bituminous shales next to the Arctic Ocean. Their erosion rate appears to be high leading to extensive slippages of the shales along the coast. Analyses have revealed that the shales contain up to 20% pyrite and 5% organic matter. Combustion occurs almost entirely in areas of slippage where a large number of fuming vents occur (Figure 3.8a). The gases contain sulphur dioxide as a result of pyrite oxidation and masks have to be worn in their vicinity. Oxidation of the shales has produced a wide variety of secondary minerals including the orange-red jarosite, a complex sulphate of iron and potassium (Figure 3.8b). Comparisons have been made between this site and Mars which also has deposits of this mineral.[19] Calculations based upon the composition of the shales indicated that the atmospheric oxidation of the pyrite and the combustion of the organic matter contribute approximately the same amount of heat to these fires.

This area is not unique in Canada. In Alberta the Smoky River is named after smouldering beds of coal on its banks.

3.6.2 The Kimmeridge Fires

A few miles west of Lulworth Cove in Dorset, UK, is the Burning Cliff of Ringstead Bay (Figure 3.9). It is a coastal landslip exposing the Kimmeridge Clay. This clay is well known for its layers of oil shale. The dark brown material has a leathery consistency which breaks up easily and readily ignites, burning with a bright yellow flame. There have been reports of the shale catching fire on a number of occasions along this coast. It was once worked to produce 'coal money' and other artwork and has a sulphur content of 6–7%. The high level of sulphur with its low self-ignition temperature of 243 °C is evidence of its easy combustibility.

The most recent Kimmeridge fires occurred in July 2000 at Clavells Hard. They seem to occur at approximately 30 year intervals. Although some reports suggest the fires were started on purpose, their origin on the steep cliffs makes this unlikely and

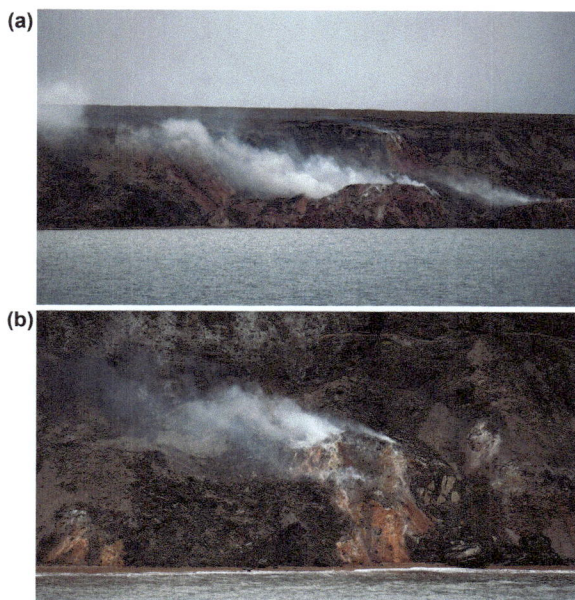

Figure 3.8 Smoking Hills, Northwest Territories, Canada. (a) General view of the burning shales, (b) close up showing red deposits of jarosite. Reproduced with permission from Colin Lang.

the southern aspect plus summer timing suggest solar warming combined with pyrite oxidation as a probable cause. Similar shales outcrop on the sea cliffs of east Yorkshire in the vicinity of Whitby although there appear to be no combustion reports from there perhaps due to their lower carbon content. Combustion has also been described from cliffs near Ballybunion in County Kerry, Ireland[20] where there was a collapse in the carbonaceous and pyrite-rich Clare Shales. They caught fire in the early 1700s.

Spontaneous combustion of these shales may well be a result of the processes described for coal. But perhaps the most likely explanation for these fires including those of Smoking Hills is frictional heat generated by landslips. These are common along the Kimmeridge coast and even a slip of a few tonnes could generate enough heat to cause ignition. Significant rock falls often generate large sparks. There are eye-witness accounts of fires started by rock falls in arid regions such as parts of South Africa for example. There have even been reports of baboons dislodging rocks on cliffs causing grass or brush fires below.

Figure 3.9 The Burning Cliff, Ringstead Bay, Dorset. Watercolour by
E. Vivian entitled *'Extent of the fire near the surface, 34 feet in
length, April 1827.'* Reproduced with permission from the
National Trust.

However this is less remarkable than an account given by
Harriet Martineau. In her *Guide to the English Lakes,* she cited
an article in the Lonsdale Magazine where a tract of dry heather
was reported to have burned without intermission for 3 weeks.
It was apparently kindled by sparks from a Cumberland cheese.
It had rolled off a cart on the road above, and bounded from
crag to crag!

3.7 LIVING VEGETATION

The 'burning bush', *Dictamnus albus* belonging to the family
Rutaceae is a small shrub that produces attractive flowers and
volatile lemon-scented oils that are easily flammable. It is
native to Eurasia and is occasionally grown in gardens. The
family Rutaceae includes many scented species including the
citrus fruits – oranges and lemons. The oils are produced by
both leaves and flowers. A cigarette lighter placed below a
flowering spike of *Dictamnus* sends a ring of yellow flames up

through the flowers without damaging them. The oils have been analysed and contain a large number of volatile compounds.[21] The most prominent is called germacrene D, a volatile sesquiterpene. The terpenes are complex hydrocarbons produced by many plant species. Germacrene possesses insecticidal properties but its combustibility has not been well studied. Some gardeners have reported spontaneous combustion in this plant but there appear to be no confirmed accounts. The oils are secreted through lysigenous cavities, mainly below the upper epidermis of the leaf and flower (Figure 3.10). *Dictamnus* is also known as the 'gas plant' and the flammable gas is mistakenly described as methane. Other strongly scented flowers worth investigating for their flammability are the Tuberose (*Heliotropium arborescens*), *Gardenia* and the moonflower (*Datura inoxia*).

Most, if not all volatile constituents of plants should be highly flammable and are known to include acetaldehyde, acetone and ethyl acetate. Acetaldehyde is probably of most interest with its auto-ignition temperature of just 175 °C. Other plant volatiles

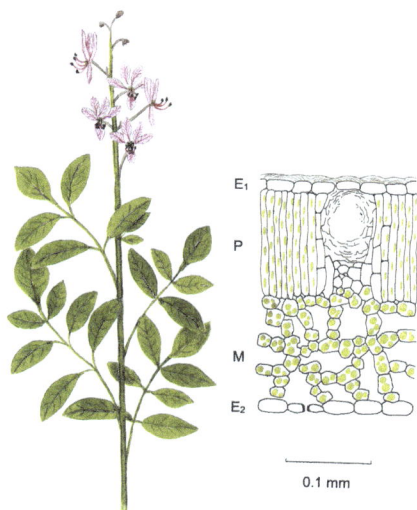

Figure 3.10   *Dictamnus albus*, the 'burning bush'. Left: A flowering stem 50 cm high. Right: Section through a leaf showing a lysigenous oil cavity among palisade cells, P. Upper epidermis with wax cuticle, E_1, lower epidermis, E_2, mesophyll cells, M.

have higher ignition temperatures but remain of interest. They include some of the low molecular weight hydrocarbons occurring in resin. Resin is highly flammable and has been previously mentioned with reference to the *pyr automaton*.

The biblical 'burning bush' might have been a *Dictamnus*. In *Exodus* 3:2 of the Old Testament, *'the angel of the Lord appeared to him (Moses) in flames of fire within a bush'* but the bush was not consumed, and God proceeded to instruct Moses from the glowing plant. The Hebrew translation of *seneh* is thought to be a reference to the bramble (*Rubus* sp.) but some scholars regard *seneh* a mistranslation of Sinai, in which case a burning mountain may have been alluded to rather than a bush. The location of the event is believed by some to be Mt Sinai. Here a monastery was built and is still surrounded by *Rubus sanctus*. This plant, a native of the region is considered to be the original 'burning bush'. *Rubus sanctus* is widespread in the Mediterranean region and resembles the common British bramble. Another biblical account occurs in Kings 1:18.38 where Elijah directs the building of two wooden pyres for the roasting of meat on Mt Carmel and challenges Baal to cause the wood to catch fire. This fails to occur, but the pyre attributed to God does so to the amazement of the Baal worshippers.

In all of these cases, excluding perhaps the last, the plants in question were presumably alive and would have had a high water content. This alone would militate against spontaneous combustion. Any local heating would be conducted away by the watery tissues and the evaporation of the water would help cool the plant. For tissues with a low water content however, plant materials can become highly combustible.[22–25] Wood auto-ignition temperatures as low as 200 °C have been reported but a figure close to 250 °C may be more realistic. These values are based upon laboratory measurements conducted with small blocks of dry wood and various methods of supplying heat. A combination of radiative and convective heat is normally used. How reliable these figures are is questionable, since ignition is dependent upon a range of factors with the size of the fragments, the kind of wood and its water content being of particular importance. The porosity of the wood, affecting the ingress of oxygen also appears to exert an influence. There is a report that softwood sawdust can ignite at 120 °C and the greater access of air to the wood particles doubtless assists in lowering the ignition temperature here. The auto-ignition temperature of paper is in the region of 220–250 °C dependent upon the nature of the product. Its large surface to volume ratio ensures a slightly lower value than that of the material from which it is usually made.

The time required for a material to self-ignite is also of interest although it has not been so well studied. Small wooden blocks, about 2.5 cm in thickness subjected to a temperature of 350–500 °C, take 100–200 seconds to flame for example. The magnitude of the supply of heat energy, known as the heat flux is often considered a better yardstick for measuring auto-ignition than temperature alone. A minimum flux for wood lies in the region of $4\ kW\,m^{-2}$.

A Canadian newspaper reported that a resident of Port Hope, Ontario, arrived home to find his house on fire. Fire crews managed to locate the origin of the fire to some decking on the second floor of the property. On one deck stood some flowers planted in plastic pots and the investigators were sure that fertiliser present on the soil had ignited them. A spokesman said that although flower pot fires were not common, they were known in the service and 2 years previously there was an almost identical incident. It was stated that the peat moss used in these pots, if left in strong sunlight, could cause fires.

Nonetheless, large pieces of wood can also ignite under some circumstances. During the construction of the Borders Railway Bridge at Berwick upon Tweed in England at about 1850, wooden piles driven being into the river bed by steam hammers were found to occasionally catch fire. This was presumably owing to the heat generated as the hammers struck.

In his History of the Peloponnesian War, Thucydides (*ca.* 410 BCE)) thought that fire could be caused by tree branches rubbing against each other. Given the well-known fire-making technique of stick rubbing, this is plausible if the branches in question were of dry, dead wood. The Arab writer Ben Mohalhal, *ca.* 942 CE also spoke of the rubbing of leaves on a cane leading to ignition.

The African baobab (*Adansonia digitata*, Bombacaeae) has been reported to self-combust. The trees often die suddenly and in the process exude an ochraceous gum[26] They sometimes collapse into a fibrous heap and in this state have been known to catch fire. Stories have circulated describing dead trees as 'vanishing overnight' or starting fires that spread to adjacent bushland. However the case for spontaneous combustion is not proven and some authorities suggest that lightning strikes are the main cause. The living baobab stores large volumes of water within its distended trunk to vitiate against drought. It is also fire-resistant and hardly a prime contender for self-ignition. Some oils have been isolated from the tree but they are not described as being excessively volatile or flammable.

3.8 ANIMALS

The conditions necessary for the combustion of animal matter would not be expected to be very different than those for plants. Those requirements being the presence of readily oxidisable material and a strong oxidant, which in our case is normally the oxygen of the air, or a substance containing it. Records of animal matter catching fire are nonetheless rare. In this section we look at two contrasting groups of animals, the insects and the mammals. The former, remarkable for their diversity and success in the biosphere, have evolved into some highly specialised forms. Among these are the bombardier beetles.

3.8.1 Bombardier Beetles

These insects have the property of producing a concentrated solution of the powerful oxidant hydrogen peroxide. These beetles are a large and successful group of insects which, when molested produce a hot spray which is expelled explosively at temperatures up to 100 °C. They have been well studied and it appears most likely that the beetles first synthesise the two reactant molecules, hydroquinone and hydrogen peroxide in separate containers.[27-28] These chemicals are then concentrated within specialised cells before being mixed for use. Upon stimulation by a predator, the mixture is passed into a reaction chamber *via* a valve where the presence of two enzymes, catalase and peroxidase, cause a sudden oxidation of the hydroquinone. The temperature rises rapidly vaporising water in the chamber. This is then expelled explosively *via* an 'exhaust' valve to startle or injure the predator. The entire process may be repeated several hundred times per second until the reactants are exhausted. Studies have shown that amphibians, feeding on the beetles frequently regurgitate them alive after having experienced the products of their armoury.

Hydrogen peroxide is produced within all living cells during the synthesis of several important molecules essential to metabolism. Because peroxide if left free in the cell can damage DNA and protein, it is rapidly removed using the enzyme catalase. In the process it is converted into free oxygen and water. Considering their high reactivity, concentrated solutions of hydroquinone and hydrogen peroxide would need specialised structures in these beetles to prevent damage to their tissues.

A study of this beetle using electron microscopy has shown that the reactants are probably formed in a group of secretory cells located high in the beetle's abdomen. The secretory system is complex with secretory cells associated with narrow ducts entering a main collecting duct (Figure 3.11). This duct is of

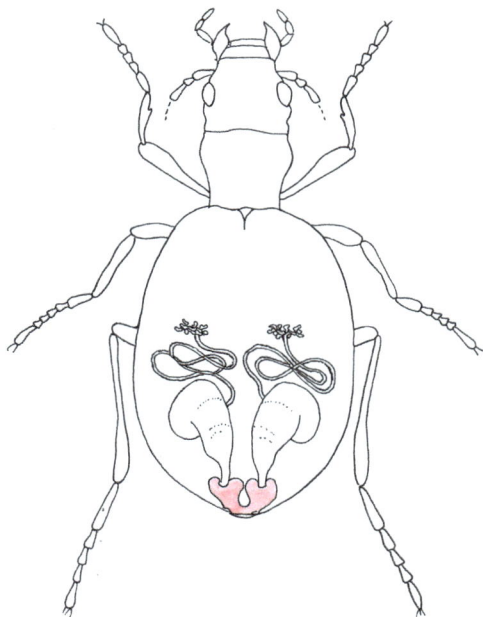

Figure 3.11 Anatomy of the defence system in the bombardier beetle *Brachinus* showing clumps of secretory lobes containing the secretory cells connected to the bladder-like reservoirs by a long collecting duct. The reactants are rapidly transferred to the reaction chamber (shown in red) when the insect is startled. Reproduced from ref. 28 with permission from Elsevier, Copyright 2015.

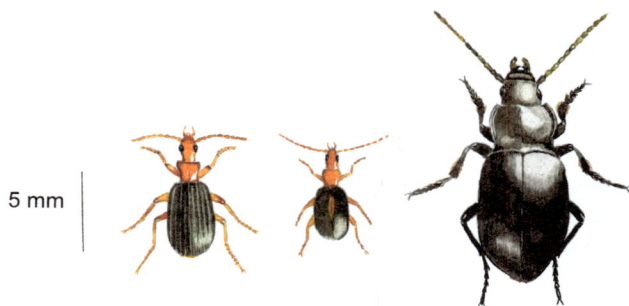

5 mm

Figure 3.12   Three species of bombardier beetle drawn to the same scale. From left to right: *Brachinus crepitatus*, *B. sclopeta* and M*etrius contractus*. *B. crepitatus* Reproduced from ref. 28 with permission from Elsevier, Copyright 2015.

unusual size, being longer than the insect itself. The reactants then enter the reservoir, a comparatively voluminous and elastic structure. Below the reservoir is a complex valve equipped with sets of muscles to control the flow of reactants. The reaction chamber itself is smaller than the reservoir, probably to allow rapid bursts of hot spray. The surface of this chamber is connected to glands that may be the source of enzymes responsible for catalysing the explosive reaction. The internal surface of the chamber is covered in long hairs, spines, or honeycomb-like sculpturing. It has been suggested that a 'paste' of enzyme may cover these structures, exposing a large surface area to the reactants thus ensuring a rapid reaction.

One of the most familiar bombardier beetles is *Brachinus crepitans* which is found in central Europe (Figure 3.12). A smaller species, *B. sclopeta,* is known from UK but is rare and confined to the London area. A particularly unusual species is *Metrius contractus*. This beetle uses a different method of defence and the hydroquinone reaction is used to produce foam to confuse its predators.[29]

3.8.2   Human Combustion

In Euripides' Greek play *Medea*, she is dismissed by Jason for a new love, Glauke. Medea takes her revenge by arranging the gift of a beautiful gown, packed in an air-tight case. Glauke opens the case, and donning the gown, twirls in front of a mirror, whence it bursts into flame and kills her. The story is depicted on vase paintings of the period. There are many more recent examples of clothing self-ignition, often with tragic consequences. There is an account of a man entering a building wearing a woollen shirt and synthetic jacket who started a fire in a carpet due to a spark from static electricity. The building had to be evacuated. When the jacket was tested it was found capable of holding a potential of 40 kV. One fire officer suggested he was just 'one step from spontaneous combustion'.

The sensationalism accruing to the apparent spontaneous combustion of human beings has hampered serious studies of the phenomenon. Human combustion is concerned mainly with the unexplained incineration of humans. Understandably it is a topic of great interest and has generated a large volume of literature over the past 200 years. Jenny Randles and Peter Hough in their book *Spontaneous Human Combustion*[30] provide an annotated list of 110 cases, mostly in Great Britain and the United States. They explored a wide range of hypotheses but noted that some of their cases had already been 'recycled' a number of times. An unbiased overview of the numerous hypotheses was attempted to explain the phenomenon but their scientific explanations are not always fully explored. Nevertheless, their accounts provide a useful sourcebook from which readers can decide for themselves whether such a phenomenon really exists.

The majority of victims appear to be unconscious or dead when combustion begins but there are a few cases where they are alive and survive. In most cases it is the clothing that is involved but there are a small number of accounts where the body itself seems to be a source of the phenomenon. When victims are already dead, a common association is an open fire into which the victim has fallen. Body fat then becomes exposed to the fire and begins to smoulder or burn, in time leading to extensive incineration. It has been suggested that the incineration results from a 'wick' effect, likened to that of a candle. The fire is continuously fed by fuel until all is consumed although several experiments to test this hypothesis have so far been inconclusive. A human body with a water content of about 70% appears an unlikely material to burn to ashes but calculations show that there is more than sufficient fuel to evaporate it. In the majority of cases large amounts of smoke are given off and some of the fat is volatilised suggesting that the combustion was slow. In closed spaces this could be due to oxygen starvation or to the nature of the material being burned. The bulk of human fat consists of long chain linoleic, oleic and palmitic acids, none of which contain more than two double bonds per molecule. It is unlikely that a fat of this composition could cause the spontaneous ignition of clothing in the manner previously described for linseed oil-soaked rags.

There are several cases where gases produced in the gut might be involved. Randles & Hough cite several individuals belching what appear to be flammable gases, in one case during a bout of gastritis. In another case, methane appeared to be issuing from a vicar's mouth while he was blowing out candles. One of the most significant observations, with clear connections to the *ignis fatuus* (Chapter 9) is a case cited by Randles and Hough. An account is given of a young man who had died of typhoid in the 1860s. Thirteen months later it was found that his coffin had split open. Workmen applied sawdust to the damage but the next day blue flames were seen issuing from the coffin. The most likely explanation is the generation of the self-igniting gas diphosphane during decay. Professor Gotfried Treviranus of the Bremen Lyceum suggested as early as 1832 that the hydrogen phosphides could be responsible for the self-ignition of humans. There is no shortage of combined phosphorus in animal tissues. The brain in particular contains significant levels and animal bones are constructed of calcium phosphate. Further discussion of diphosphane is provided in Chapter 9.

Many other suggestions have been put forward to account for human combustion, but none appear to have been conclusively demonstrated. They include the ignition of greasy human hair by water droplets focused by sunlight and the electrolysis of water within the body producing hydrogen and oxygen. In this last case there would still need to be a source of ignition. A BBC programme in November 2013 made reference to spontaneous human combustion and reported a recent account in Galway. A spokesman noted that *'a fire in the fireplace of the sitting room had not been the cause of the blaze.' Only the body showed signs of burning'*. It was mentioned that ill people produce some acetone, the flammable vapours of which could be responsible for the phenomenon.

Randles and Hough also mention the apparent rarity of animal self-combustion. One case they reported was that communicated by Mr. R. Reed who described an incident on the downs near

Weymouth during WW2. On a dark night, a fire was seen to erupt about 100 m distant. On closer inspection was found to be a sheep on its side with blue flames issuing from its stomach. The flames flared up yellow and caught alight to its fleece. The authors suggest that few such cases might be reported because farmers would blame the occurrence on a lightning strike. For if another explanation was given, it would be more difficult to claim the insurance. There have also been reports of self-ignition in dogs and pigs from time to time but as is often the case, hard evidence is lacking.

FURTHER READING

P. C. Bowes, *Self-heating. Evaluating and controlling the hazards*, Elsevier, Amsterdam, 1984.

M. Harrison, *Fire from Heaven*, Pan Books, London, 1977.

A. Mayor, *Greek Fire, poison arrows and scorpion bombs. Biological and Chemical Warfare in the Ancient World*, Duckworth, Woodstock, UK, 2003.

J. R. Partington, *The history of Greek Fire and Gunpowder*, Johns Hopkins Press, Baltimore, Maryland, 1960.

H. Rossotti, *Fire: servant, scourge and enigma*, Oxford University Press, 1993.

H. P. Rothbaum, Spontaneous combustion of hay. *J. Appl. Chem.*, 1963, **13**, 291–302.

REFERENCES

1. A. Egri, A. Horvath, G. Kriska and G. Horvath, Optics of sunlit water drops on leaves: conditions under which sunburn is possible, *New Phytol.*, 2010, **185**, 979–987.
2. G. A. Cowan, A natural fission reactor, *Sci. Am.*, 1976, **235**, 36–47.
3. R. Buggeln and R. Rynk, Self-heating in yard trimmings: conditions leading to spontaneous combustion, *Compost Sci. Util.*, 2002, **10**, 162–182.
4. H. R. Goeppert, *Über die warm-entwicklung in den Pflanzen, deren gefreiren und die Schutzmirtel gegen dasselbe*, Breslau, 1830.
5. J. S. Haldane and R. H. Makgill, Spontaneous combustion of hay, *Fuel Sci.*, 1923, **11**, 380–387.

6. H. Ranke, Experimentelle beweis der Moeglichkeit der selbstentzuendung des Heues (Grummets), *Liebigs Ann.*, 1873, **167**, 361–368.
7. C. A. Browne, *The spontaneous combustion of hay*, U.S. Department of Agriculture, Technical Bulletin, Washington, 1929, vol. 141.
8. G. Laupper, Die Neuesten Ergebnisse der Heubrandforschung, *Landw. Jahrb. Scheiz. Jahrg.*, 1924, **34**, 1–54.
9. M. Westram *et al.*, *Dust explosion in a sugar silo tower: investigation and lessons learnt*, Symposium Series, Crown Copyright, UK, 2008, vol. 154.
10. P. J. Wakelyn and S. E. Hughes, Evaluation of flammability of cotton bales, *Fire Mater.*, 2002, **26**, 185–189.
11. C. J. Abraham, A solution to spontaneous combustion in linseed oil formulations, *Polym. Degrad. Stab.*, 1996, **54**, 157–166.
12. R. Wetherill, *The Ancient Port of Whitby and Its Shipping*, Horne & Son, Whitby, 1905.
13. H. N. Khoury, Long-term analogue of carbonation in travertine Uleim Quarries, Central Jordan, *Environ. Earth Sci.*, 2012, **65**, 1909–1916.
14. T. Pennant, *A Tour of Scotland, MDCCLIX*, 2nd edn, London, 1773.
15. S. C. Banerjee, *Spontaneous combustion of coal and mine fires*, A.A. Balkema, Rotterdam, 1985.
16. G. B. Y. Christie and D. E. Mainwaring, Oxidative and immersional heating on low rank coal surfaces, *Fuel*, 1992, **71**, 443–447.
17. J. S. Haldane, Spontaneous firing of coal, *Trans. Inst. Mine Engnrs.*, 1916–17, **53**, 194–230.
18. G. Martinelli, S. Cremonini, E. Samonali and G. B. Stracher, Italian Peat and Coal fires, in *Coal and Peat Fires: a Global Perspective*, ed. G. B. Stracher, A. Praaksh and E. V. Sokol, Geology and Combustion, Elsevier, Amsterdam, 2015, vol. 1, pp. 40–73.
19. S. E. Grasby and J. B. Percival, *et al.*, Extensive jarosite deposits formed through auto-combustion and weathering of pyritiferous mudstone, Smoking hills, Northwest Territories, Canadian Arctic – a potential Mars analogue, *Chem. Geol.*, 2022, DOI: 10.1016/j.chemgeo.2021.120634.

20. C. Smith, *Ancient and Present State of the County Kerry*, Privately published, 1756.
21. D. Velickovic, Chemical analysis of *Dictamnus albus* essential oil from Serbia, *Agro Food Ind. Hi-Tech*, 2012, **23**, 4–6.
22. V. Babrouskas, Ignition of wood. A review, in *Interflam 2001*, Interscience Communications Ltd, London, 2001, pp. 71–88.
23. F. G. Bell and L. J. Donnelly, The problem of spontaneous combustion illustrated by two case histories, in *9th Congress of the International Association for Engineering Geology and the Environment*, Durban, South Africa, Sept. 16–20, 2002.
24. P. de V. Booysen and N. M. Tainton, *Ecological Effects of fire in South African Ecosystems*, Springer-Verlag, Berlin, 1984.
25. H. Kubler, Heat generating processes as cause of spontaneous ignition of forest products, *For. Prod. Abstr.*, 1987, **10**, 299–327.
26. G. E. Wickens and P. Lowe, *The Baobabs*, Springer, 2008.
27. J. Dean, D. J. Aneshansley, H. E. Edgerton and T. Eisner, Defensive spray of the Bombardier Beetle: A Biological Pulse Jet, *Science*, 1990, **248**(4960), 1219–1221.
28. A. Di Giulio, M. Muzzi and R. Romani, Functional anatomy of the explosive defensive system of bombardier beetles (Coleoptera, Carabidae, Brachininae), *Arthropod Struct. Dev.*, 2015, **44**, 468–490.
29. T. Eisner, D. J. Aneshansley, M. Eisner, A. B. Attygalle, D. W. Alsop and J. Mainwald, Spray mechanism of the most primitive bombardier beetle (*Metrius contractus*), *J. Exp. Biol.*, 2000, **203**, 1265–1275.
30. J. Randles and P. Hough, *Spontaneous Human Combustion*, Robert Hale, London, 1992.

CHAPTER 4

St Elmo's Fire and Related Electrical Phenomena

4.1 INTRODUCTION

The corona discharge called St Elmo's fire, also known as the *ignis lambens* (licking fire), is an electrical phenomenon responsible for emitting light from exposed objects such as aircraft wings, lightning conductors, ship's masts (Figure 4.1) and other tall or sharp objects. The origin of the name St Elmo's Fire is obscure. It may have been derived from St. Erasmus, the patron saint of sailors in the Mediterranean where its presence was believed to reveal his benevolence toward them. Many other names have been applied such as St Nicholas, St Peter and St Helmes fire. Coronas attached to horses' manes and men's heads were called *haggs* and presumably much better known in the days of non-motorised transport.

The emission of light is caused by fast electrons interacting with gases in the atmosphere and is therefore a type of electroluminescence. Corona discharge involves the passage of electricity through gases and before looking at the phenomenon in more detail it will be useful to briefly review the history and nature of electricity itself.

Luminous Phenomena: A Story of Spontaneous Combustion, Phosphorescence and Other Cold Lights
By Allan Pentecost
© Allan Pentecost 2025
Published by the Royal Society of Chemistry, www.rsc.org

Figure 4.1 St Elmo's Fire seen as 'seven lighted candles' on a ship's mast followed by storms on Christopher Columbus' second voyage across the Atlantic, 1492–1493. Reproduced from L. Figuier's *Merveilles de la Science* of 1859.

4.2 HISTORY AND NATURE OF ELECTRICITY

Thales of Miletus (*ca.* 600 BCE) noted that pieces of amber (known to the Greeks as *elektron*), when vigorously rubbed, will attract light objects such as straw or dried grass. It was not until late in the Renaissance in 1600 that William Gilbert of Colchester wrote a seminal review on electricity and magnetism. It inspired later scientists to make important electrical discoveries. In 1660 Otto von Guericke of Magdeburg built the first successful electrical machine. A large ball of sulphur was rotated mechanically and a means of friction applied. Large quantities of electricity were produced giving small sparks and electric shocks. With the rapid dissemination of knowledge through advances in printing, news spread rapidly throughout the continent, and more advanced machines soon followed. Stephen Gray of London reported experiments in the early 1700s where the warming or rubbing of many other materials caused them to electrify. He found that electrified linen and paper could emit light in the dark. A pointed iron rod suspended by silk gave off cones of light in the dark when it was approached by an electrified piece of glass. At around the same time, Charles du Fay of France concluded that there were two kinds of electricity, namely 'vitreous' produced on glass, animal hair and quartz, and 'resinous' produced from amber, thread and paper. Benjamin Franklin concluded that electricity was a fluid contained in all bodies. When at equilibrium, the fluid contained was neutral but if in excess the body was deemed positively charged, and when wanting it was negatively charged. Vitreous electricity was regarded as positive. In 1753 Franklin suspended electrically charged bodies in rosin smoke and observed beautiful patterns caused by the strong electric field.

It is now known that static electricity consists of electric charges that become dispersed over solid or liquid surfaces. Negatively charged electrons exert a strong repulsive force upon one another, and this force varies with the distance between them. The force is such that electrons close together repel one another much more strongly than those far apart. This repulsion has been shown to follow a simple relationship called the Law of Force. The movement of free electrons results in an electric current, but electrons easily become attached to atoms and molecules to produce ions. Ions are also charged but are not as mobile as electrons.

A positive charge occurs on the proton which sits in the nucleus of an atom. Protons are so strongly bound that positive ions normally result from the loss of an electron from an atom or molecule rather than the gain of a proton. Positive ions are much heavier than electrons and are less mobile when present in a gas. Positive charges also repel one another but are attracted to a negative charge. This fundamental difference between the two types of charge goes some way towards explaining electrical phenomena – most of the 'movement' of charge is undertaken by the mobile electron. In gases and liquids, positive charge does have some mobility but it is several hundred times less than that of the electron. Since atoms are electrically neutral some process must be responsible for separating positive and negative charges to produce electrical phenomena. Friction is clearly one of these.

An electric field is a region of space where charged particles experience a force. The fields are not visible but can be understood by mapping out 'lines of force'. At any point on a line of force, the direction of the force is in the direction of that line. The force may be in either direction along the line dependent upon whether it is repulsive or attractive, as for example between an electron and a positive charge. To avoid confusion, lines of force are equipped with arrows showing the direction of force towards a positive charge. Since the force varies according to the amount of charge and the distance between the charges, some means of defining the intensity of the force is needed. The unit of electrical intensity is the volt per metre.

Lines of force assist the understanding of variation of electric intensity at any point in space. High and low intensities are represented by a high and a low density of lines of force. The number of lines of force per unit area (*e.g.* per square centimetre or square metre) is termed the flux. Imagine an isolated expandable sphere 5 cm across and place upon its surface a fixed number of charges, either positive or negative. The charges are able to move around on the surface in response to nearby charges, but there will be no net flow of charge away from the sphere's surface (Figure 4.2). Each charge is provided with one line of force. Because like charges repel, they will be distributed equally over the surface. An electric field is associated with these charges shown by the radiating lines of force as if all the charge were concentrated at the sphere's centre. The sphere is now expanded to five times its original surface area while the number of charges remains the same. The charges are now spaced five times further apart and the electric field is reduced in intensity. Its total flux, equal to the number of lines of force, remains the same. It is clear that the charge density has changed and is reduced as the surface area increases. Likewise, as the sphere is shrunk its charge density will increase. It follows that small spherical objects are surrounded by much stronger electric fields than larger ones carrying the same charge. A theorem attributed to Carl Gauss shows that any closed surface independent of its shape gives the same result. It means that small pointed objects can carry many charges associated with strong electric fields. This is even true if the pointed object is part of a larger structure.

Air is a poor conductor of electricity but strong electric fields above negatively charged projections cause the air molecules to respond by induction and a region of positive charge can develop above them. Some of the negative charges (electrons) are lost to the air as incoming positive charges start to bombard the surface. Unless there is a continuous flow of electrons to the point the charges are eventually neutralised. This process is known as

(a)

(b)

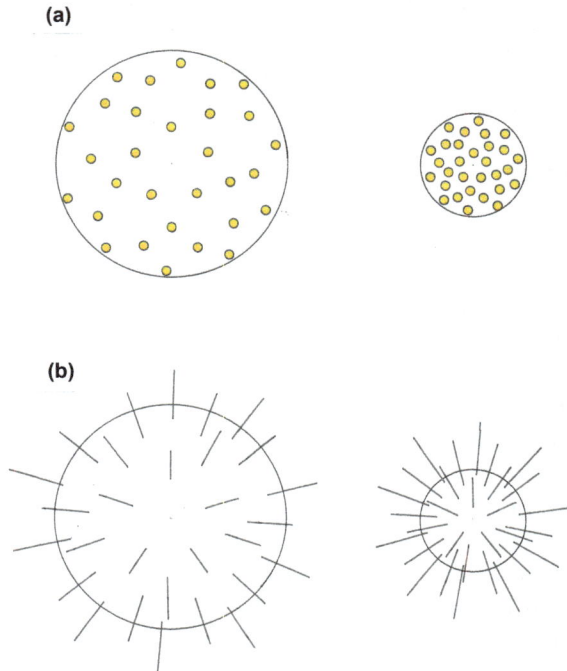

Figure 4.2   Diagram of an expandable sphere with surface electric charges. In (a) the charges (yellow circles) arrange themselves equidistant from each other. As the diameter of the sphere is reduced, the charges remain and the electric field strength (b), shown as one line per charge increases.

'charge leakage' and can cause serious losses in high voltage cables. Loss of charge on pointed conductors is easy to demonstrate experimentally using an electroscope. Corona discharge or St Elmo's Fire is a manifestation of this phenomenon.

4.3   SOME ACCOUNTS

Corona discharges were known long before they were understood. There is an account by Pliny in his *Natural History* where he describes the ability of St Elmo's Fire to pass from one object to another. Their appearance as pairs (Castor and Pollux) was a good omen aboard ship, and sailors used to interpret the future weather according to the form of the light. Pliny described it as *stars [that] settle on the masts and on the heads of men* and it was known from the tips of soldier's spears according to Livy. Perhaps surprisingly, of the 500 or so accounts of fire in the Bible, few seem to relate to the phenomenon. Tongues of fire are mentioned in Acts 2:3 and there is the Burning Bush of Exodus 3:2. Pillars of fire are mentioned in Exodus 13:21 that appear at night as a guide and there are references to either St Elmo's or lightning in Exodus 9:24 and Numbers 9:15–23. The numerous fire references in the Quran do not appear to relate to it. Despite being recognised for millennia, there is little folk-lore associated with it.

In *Moby Dick*, the seafaring adventure by Herman Melville, the whaling ship *Pequod* encounters a typhoon off Japan where thunder, lightning and displays of St Elmo's fire are described:

> *"The rods! The rods!" Cried Starbuck to the crew. "Look aloft! The St. Elmo's Lights, corposants!"*
>
> *All the yardarms were tipped with a pallid fire and touched at each tri-pointed lighting-rod end with three tapering white flames, each of the three tall masts was silently burning in that sulphurous air, like three gigantic wax tapers before an altar.*
>
> *While this pallidness was burning aloft, few words were heard from the enchanted crew; who in one thick cluster stood on the forecastle, all their eyes gleaming in that pale phosphorescence, like a faraway constellation of stars.*
>
> *"Aye, aye men!" cried Ahab. "Look up at it; mark it well; the white flame but lights the way to the White Whale! Then turning- he put his foot upon the Parsee and with fixed upward eye, and a high-flung right arm, he stood erect before the lofty tri-pointed trinity of flames.*
>
> *"The boat, the boat!" cried Starbuck, "look at thy boat, old man!" Ahab's harpoon remained firmly lashed in its conspicuous crotch, so that it projected beyond his whaleboat's bow; but the sea that had stove in its bottom had caused the loose leather sheath to drop off, and from the keen steel barb there now came a levelled flame of pale, forked fire. As the harpoon burned there like a serpent's tongue, Starbuck grasped Ahab by the arm "God, god is against thee, old man; forbear! 'Tis an ill voyage! Ill begun, ill continued".*

Herman Melville had a remarkable life and spent time on whaling boats in the late 1830s. His description of St Elmo's fire leaves little doubt that it was based upon an actual sighting.

Most accounts of St Elmo's Fire refer to sea-going vessels, where formerly look-outs were posted at night, often on the ship's masts. Antonio Pigafetta kept a journal of Magellan's voyage to South America in the *Trinidada* where they encountered great tempests: *Water was driven in level sheets and sputtering lights appeared at the mastheads, the yardarms where pale flames streamed. A ball of fire hung at the flagship's mast tops, clinging in the wind and rain.* He goes on to state that *the body of St Anselm appeared to us several times. One night it was very dark on account of the bad weather the saint appeared in the form of a fire lighted at the summit of the mainmast and remained there near two hours and a half, which comforted us greatly, for we were in tears only expecting the hour of perishing. And when that holy light was going away from us, it gave out a great brilliancy in the eyes of each, that we were like people blinded and calling out for mercy. For without any doubt nobody hoped to escape that storm.*

There are a number of accounts related to corona discharges on or near mountains. In the 1949 volume of the British journal *Weather*, two climbers were described descending the An-t-Sron Ridge of the mountain of Bidean nam Bean at Glencoe, Scotland.[1] It was late December and one of the climbers was seen to have acquired a green-white glow caused by brush discharges from particles of ice on his balaclava, visible from 5 metres away. Other brush discharges were seen on the tips of their mittens and later their ice axes began to sizzle and emit fine purple brushes 2 centimetres or more in length. They ceased once the two axes were brought close to one another. They also ceased when the axes were held out of the wind. The display lasted for about 15 minutes. There was no thunderstorm activity but a strong wind blowing over the ridge picked up ice particles the size of small hail and they were considered the likely cause. The climbers even found they could 'spit fire' like a dragon.

In *Scrambles amongst the Alps*,[2] the mountaineer Edward Whymper recalled being uncomfortable when hearing mysterious rushing sounds while close to the summit of the Théodule Pass in Switzerland during stormy weather. It sounded as if a sudden gust of wind was sweeping along the snow yet the air was perfectly still. There was, he noted, a similar encounter there in 1842 described by Principal Forbes who said that his fingers 'yielded a fizzing sound' and yet another by Spence Watson on the Aletsch Glacier. He reported that his hair stood on end in an uncomfortable but amusing manner and the veil of another member of the party stood upright in the air. In the *Alpine Journal*, Robert Watson[3] reported pricking and burning sensations during an alpine storm which ceased when there was a clap of thunder. There are also a number of anecdotal reports describing 'mountain glows' particularly from the Andes. These may represent large corona discharges but they have been little researched.

Cade and Davis[4] relate a further story of a young girl walking her bicycle up the steep hill on Hawkshead Moor in the English Lake District. A thunderstorm was approaching and as she gained the summit every lightning flash was followed by blue flames of St Elmo's Fire running across the handlebars. She cycled down the hill as fast as she could.

Aircraft are subject to corona discharges and today, this is where they are most likely to be seen. Tongues of light up to 40 cm long can appear on propeller arcs. The tips of rotating helicopter blades can also show them, sometimes in a spectacular manner. Aircraft receive an electric charge through a range of processes. Ice particles from cirrus clouds are particularly effective. As an ice crystal impacts upon the aircraft surface, there is a separation of charge, with a negative charge passing to the craft while a positive charge remains on the rebounding crystal. Particles of dust, volcanic ash, rain drops and even air molecules can behave similarly. It is analogous to the static electricity generated by rubbing amber with a dry cloth. Aircraft can also be charged in the strong electric fields that sometimes occur between clouds. Once the potential rises to around 100 kV per metre (1 kV = 1000 volts), corona discharges can occur on sharp edges and points such as propellers and wing tips, wing edges and aerials. Since the 1930s when radios were first used on aircraft, there has been a concerted effort to reduce these discharges as they interfere with radio transmission. The effect reaches extremes on spacecraft re-entering the earth's atmosphere where the interference is so great that radio communication is lost for several minutes. Many devices have been used to counteract the effect in aircraft, one of the most effective being discharge wicks. These are narrow metal-clad cylinders containing hundreds of exposed carbon fibres with pointed ends. They allow the excess charge to flow back into the air *via* microscopic discharge points situated on the fibres. Under most situations these wicks prevent the aircraft charging up sufficiently to give corona discharges. Wicks can be seen on the trailing edges of modern aircraft wings such as those of the Boeing 737 and the Airbus 380. They do not eliminate static altogether and earthing of the aircraft is still required after landing.

The Hindenburg disaster of 1937 has already been mentioned in connection with hydrogen combustion. One account of the incident noted that about a minute before the fire was discovered, St Elmos Fire could be seen flickering along the airship's back. This observer was fortunate in viewing the ship against a backdrop of darkening skies as the discharge is not usually seen in daytime. The energy density of a corona discharge is much lower than that of an electric spark and unlikely to be the cause of the ignition of the Hindenburg. It has also been suggested that the flickering might have been caused by hydrogen since it burns with a faint blue flame.

Then there is the mysterious fire starting in York Minster after what appeared to be a type of corona discharge, preceding a lightning strike. This was reported about 2.30 am on July 9th, 1984, by observers who claimed to see the strike. Lights had been seen on the rooftops several hours before, though no thunderstorm activity in the immediate area was recorded that day. The resulting fire caused much damage even though the building was protected with lightning conductors. Others reported seeing a 'dark mass' over the church from which orange or yellow bolts of electricity arose, striking the church. This might have resulted from a powerful 'bolt out of the blue' overloading the conductor thus starting the fire.

Corona discharges are sometimes reported around high tension electricity lines and their pylons. Exceptionally tall masts such as that on Winter Hill, Lancashire (309 m) and the Arqiva Tower on Emley Moor, West Yorkshire (330 m) had reports of nocturnal lights in the 1970s and 80s. The origin of these lights is still unclear since upon such tall structures corona discharges would be difficult to see from the ground.

There may also be connections with the mysterious Owl Man. This story seems to have started with the sighting of a huge owl in Mawnan Wood reported in the *Cornwall Chronicle* of 1926. It was followed 50 years later by an incident that occurred around Mawnan Church in April 1976 when two young children saw an object hovering over it. They described it as a partly feathered man-sized owl that was accompanied by a hissing sound. The church sits close to the coast above a 50 m cliff and would be in a good position for corona discharge. Another possible cause is an insect swarm. Laboratory studies have shown that flying insects can glow under a strong electric field emitting sufficient light to be visible at night.[5]

4.4 POSITIVE AND NEGATIVE CORONAS

It has already been noted that corona discharges occur on sharp points in the strong electric fields that are often associated with thunderstorms and tornadoes. The form of this discharge varies according to the polarity of the field. When the point is positively charged the light is more focussed and produces a diffuse purplish glow, whereas a negatively charged point emits an approximately spherical reddish glow up to the size of an orange, often accompanied by brush discharges – greenish or violet streamers about 10 cm long that may emit a hissing or cracking sound. Positive fire appears to be observed more frequently. This might be related to the observation that lightning strikes most frequently transfer negative charge to the ground, although this type of corona discharge also tends to be more conspicuous. The phenomenon is most often reported on good conductors such as metals but this is not always the case as we have already noted. It can also occur on water drops and on ice, a poorly conducting material. On ice the discharge tends to be weak and intermittent owing to the restricted movement of charge across its surface. However, an interesting account of a possible *ignis lambens* is given by W. Ganong on the frozen Baie des Chaleurs, New Brunswick.[6] In describing the accounts he noted that the large light could often be seen at night before a storm, even when the ice had formed. It was likened to the 'flame-lit rigging of a ship'. The *ignis lambens* was sometimes confused with the *ignis fatuus* to be described later. Despite the name, combustion is not involved in the emission.

4.5   THE POSITIVE CORONA

In the case of both positive and negative coronas, a pointed conductor will be subject to environmental radiation in its vicinity due to the universal presence of cosmic rays and ultraviolet photons in the atmosphere. These may be of sufficient energy to fragment a neutral air molecule into a positive ion and electron. Since the conductor is already highly charged as a result of atmospheric electricity the electron will be accelerated towards it if the conductor is charged positive (positive corona). In the process this electron will collide with other air molecules as it forces its way towards the point. During these violent collisions more ions are generated and the charges are again separated by the strong electric field. Further electrons rush towards the point only to encounter yet more air molecules producing an 'electron avalanche'. This results in a small region of space close to the charged point containing a large number of ions producing a plasma. Both positive ions and electrons gain a considerable amount of energy in the electric field and during recombination this excess energy is lost by the expulsion of a photon of radiation (Figure 4.3).

Some of this radiation is in the ultraviolet region and further electrons are released from atoms surrounding the plasma through UV photon interactions. These electrons are also drawn towards the point by electrostatic attraction. The actual proportion of ionised species is small when compared with the neutral air molecules but there are also a large number of excited neutral molecules present, particularly nitrogen that can also emit light during de-excitation. At the edge of the corona the electric field is too weak to produce plasma. Positive coronas differ from negative coronas because the electrons are attracted in opposite directions.[7] They tend to be smaller in size and are usually bluish-white in colour (Figures 4.4 and 4.5). They have a lower electron density than negative coronas and generate less ozone and nitrogen oxides.

4.6   THE NEGATIVE CORONA

Their mechanism initially follows the same type of chain reaction as the positive corona. But in this case the resulting electrons are repelled away from the point since it is negatively

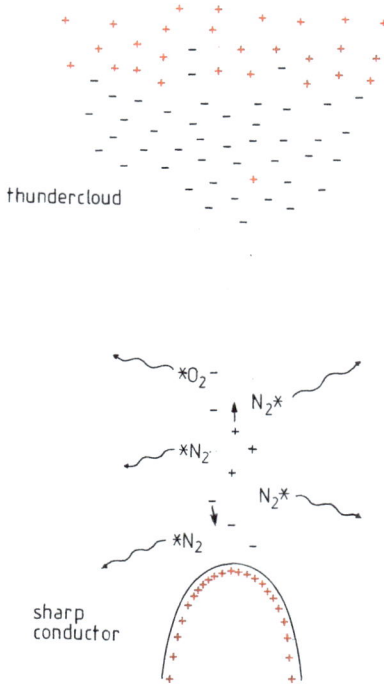

Figure 4.3 Diagram of a positive corona. A thundercloud generates a nega-tive charge in its lower region, inducing an intense positive charge on sharp points. Environmental radiation ionises the air near the points starting an electron avalanche. Collisions within the resulting plasma generate excited molecules and ions pro-ducing visible radiation of the discharge (arrows with curved tails).

charged. They soon lose energy as the electric field strength falls away. In the resulting plasma some of the more energetic pho-tons generated during ion recombination strike the point. Here they can detach further electrons on the surface of the point as they are more easily liberated than those within air molecules. In addition, electrons are liberated by positive ions striking the point since they are attracted towards the negative charges. The resulting discharge tends to be more reddish in colour and may extend as a series of points along the tip.

Within a negative corona discharge the electron cascades can show pulsing behaviour. As the dislodged electrons move away from the pointed surface they slow down to such an extent that

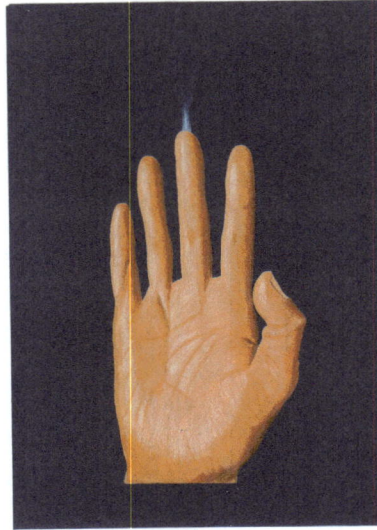

Figure 4.4   St Elmo's fire as it appears on a person's hand. The middle finger receives most of the charge and the air above emits a cone of bluish-white light caused by excitation of nitrogen molecules as electrons are attracted to the tip. Above this area is a faint brush-like emission of violet light along with some UV radiation resulting from photoluminescence (positive corona).

Figure 4.5   Corona discharge on spruce needles. (A) Positive fire, (B) negative fire. Reproduced from ref. 7 with permission from Elsevier, Copyright 2022.

they are no longer able to ionise the air. Instead, they are captured by oxygen molecules producing O_2^- ions. As the pulse proceeds the cloud of negative oxygen ions increases in size. Meanwhile positive ions are drawn to the point by electrostatic attraction where they are dispersed, temporarily reducing the strength of the electric field. The field then builds up again and the process is repeated. This can cause a regular pulsation in the electric field and consequent light emission. The reddish to purple glow of the negative corona is caused by light emission from the excited nitrogen molecules.

Both positive and negative corona discharges are often accompanied by faint fan-shaped plasma streamers. These are known as brush discharges and tend to occur on broader points, 5 mm or more wide. They are related to electric sparks but they cease a short distance into the air and tend to be illuminated only at their tips.

Corona discharge occurs in air when the surface charge density reaches around 2.7×10^{-5} coulombs per square metre. Since the electron has a charge of 1.6×10^{-19} coulombs, this is equivalent to rather more than 10^{14} free electrons per square metre of surface. This is a large number of charges, but with a surface containing around 2×10^{19} atoms per square metre, only about eight atoms per million would be needed to accommodate the additional charge.

Corona discharge is a serious source of power loss in high voltage lines and much is done to try and reduce it. Metallic 'corona rings' are sometimes attached to lines where sharp angles cannot be avoided. To avoid leakage in high potential transmission lines, the cables are sometimes made hollow to reduce the line's radius of curvature and thus reduce the electric field strength and the leakage of current. Corona discharge also has many useful practical applications. For example ozone is often produced in air subject to intense electric fields and this is put to good use in the ozonisers used for water purification.

4.7 SPARKLING MERCURY

The French astronomer Jean Picard was used to moving his barometers from place to place. He discovered that in the dark, an illuminated band appeared at the mercury meniscus when the mercury moved down the tube. This observation was taken up by Francis Hawksbee. He found that light was only emitted by mercury globules that rolled down the sides of glass. It was particularly noticeable if the surrounding air was partially evacuated. Hawksbee concluded that the light was emitted as a result of the rubbing of the metal against the glass. He compared the small bright flashes with those of lightning and his observations with other light-emitting materials earned him the title of the founder of electroluminescence. A role for electricity was confirmed by Christian Ludoff, who in 1745 found that silk threads would move in response to the flow of mercury over glass.

More recent work by R. Budakian and colleagues[8] showed that as mercury rolls within a rotating glass tube its movement is not smooth but jerky. They termed the motion stick-slip friction and noticed that electrical discharges were involved. Electrons originating from the mercury surface attach to the adjacent glass producing a static charge sufficient to pull it along until an electrical discharge takes place, which then prevents its flow. The electric field is strong enough to accelerate electrons to sufficient energies to ionise the surrounding gas followed by excitation and light emission. Such stop-start phenomena had not been observed before and occurred on a very short time scale. As the gas pressure within the rotating tube was reduced, light emission was found to be similar to that of a corona discharge.

If the gas surrounding the mercury is a strong emitter in the visible region the results can be impressive. Formerly tubes containing mercury and neon gas at low pressure were made in the form of earrings giving bright red flashes of light at the turn of a head. Stick-slip friction accompanied by light emission can also be seen if Scotch tape is pulled off its reel in the dark. Although considered a form of triboluminescence, static discharge is also likely to occur.

4.8   CORONA DISCHARGE PHOTOGRAPHY AND PYROGRAPHY

In recent years an art form known as corona discharge photography has made an appearance. An earthed metallic plate is placed upon a suitable insulator and covered with a small sheet

Figure 4.6 An example of pyrography made on a piece of softwood furniture. Image *ca.* 40 cm across.

of glass. Photographic film is placed on top and finally an object that is to serve as the subject of the art form. Popular objects have been leaves or similar flat shapes. The object is then connected briefly to a high voltage/high frequency generator causing small corona discharges to appear around the edge of the object. These create intricately branched streamers that are faithfully recorded on the developed film. The patterns are known as Lichtenberg figures after their discoverer, Georg Lichtenberg. Similar figures can be produced on a larger scale using higher electric currents. A plane block of wood covered with a layer of electrolyte and connected to high voltage terminals produces an intricate pattern as the current moves through the surface of the wood, burning it as it goes. The process is known as pyrography but it is a hazardous operation owing to the high voltages employed and the potential risk of fire (Figure 4.6).

4.9 LIGHTNING

Earth's surface usually has a weak negative electrical charge and the ionosphere far above the earth a positive charge leading to a potential difference of about 100 volts per metre. Sparking cannot occur under these conditions because air is a poor conductor of electricity. It takes exceptional conditions to alter this situation. Within large clouds however, complex interactions between the condensation of water vapour, its accumulation and freezing into ice, and the gravitational pull upon the ice by the earth combine to increase the electric charge difference dramatically.

It is well known that the condensation of water vapour into liquid releases heat causing the expansion of the surrounding air. Expansion makes the air buoyant allowing it to rise. As it rises it cools through the adiabatic effect leading to further condensation and expansion. This process can continue long enough to allow clouds to reach great heights, well over 7 km above the earth's surface. Within these high clouds there are strong air flows in all directions and they help the water droplets to coalesce. At a sufficient height they freeze to form ice crystals or larger aggregates known as soft hail or graupel.[9] Although the precise mechanism is not understood, graupel receives an electrical charge that is temperature dependent. It probably results from its collision with other ice particles. Below about −15 °C graupel is negatively charged while above that temperature it becomes positively charged. Thus if there is a downdraft in a region where the cloud is below −15 °C large amounts of the negatively charged ice will descend to a lower level bringing negative charge with it. In other parts of the cloud, or in an adjacent cloud, the ice may retain a positive charge. These accumulations and movements of electric charge lead to huge potential differences of the order of 20 kV or more per metre. This can occur between different areas of the same cloud, between different clouds or between the cloud and the earth's surface. Under the influence of such intense electric fields the atoms and molecules that make up the air start to break down. Although air atoms are neutral, their electrons are influenced by the strong electric field. In addition, 'environmental radiation' consisting of charged particles is invariably present as noted with the corona discharge. At some critical field strength the air breaks down to form a plasma and a spark occurs. The initial spark, known as a leader, heats up the air to a sufficiently high temperature to ionise some of the air molecules. This makes a conducting channel allowing a large amount of current to flow in a return stroke giving the familiar lightning bolt.

The brilliant flash of lighting occurs in the first few microseconds of the return stroke[10] (Figure 4.7). Temperature rises to

Figure 4.7   A cloud-to-ground lightning strike at Aston, England. The split-
ting of lightning bolts into irregular branches is influenced by the
physical and chemical heterogeneity of the atmosphere. Image by
Steve Dalgliesh.

20 000 °C or more and the gas in the lightning channel is turned
into a plasma consisting of highly excited N^+ and O^+ ions. De-
excitation leads to very strong emission lines that light up the
sky. Plasma has been described as the fourth state of matter. It is
not so clearly defined as the three more familiar states, solid,
liquid and gas but it is the most common state in the known
universe. Because it consists of charged particles it is influenced
by electric and magnetic fields. Since lightning is accompanied
by heat the light can be described as incandescent but perhaps
more satisfactorily as electro-incandescent.

Detailed studies of lightning over many years have established
that about 90% of the cloud-to-ground flashes transfer negative
charge to the earth with the remaining 10% transferring positive
charge. The distribution and frequency of lightning flashes have
also been studied and detailed maps have been prepared. Flash
density, the number of flashes per square kilometre per year,
falls from the poles to the equator. Polar regions experience
densities of around 0.01 compared with 30–50 for the moist
tropics. In Europe the density ranges from less than one in
the north to about ten in the south. Globally there are about
100 flashes every second. Since the majority of air-ground strikes
are negative there must be some means of returning the charge
to the atmosphere to maintain electrical neutrality. There is
known to be a slow and continuous leak of charge from the earth

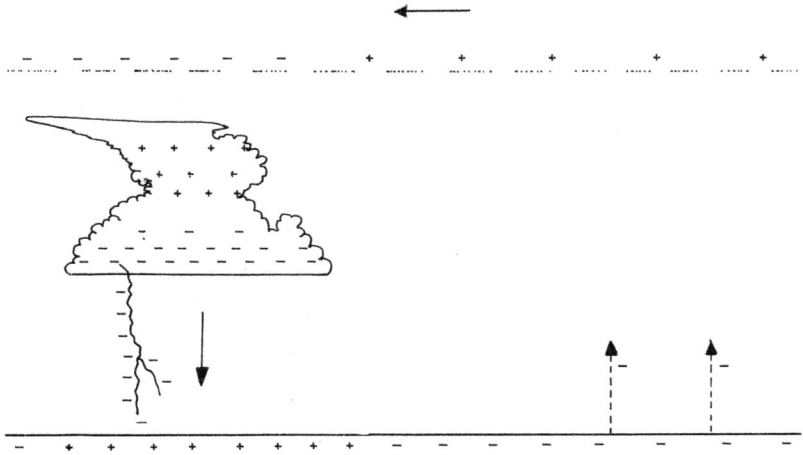

Figure 4.8   The global electric circuit. A lightning strike conveys a negative
charge to the earth, restored further afield by charge leakage back
into the atmosphere. The circuit is completed with the positively
charged ionosphere (upper arrow).

back into the atmosphere *via* conduction. This occurs under 'fair
weather' conditions away from thunderstorms and the atmospheric conductivity is provided by environmental radioactivity
and cosmic rays as mentioned above. The circuit is completed by
another conducting region high in the atmosphere known as the
electrosphere or ionosphere (Figure 4.8).

It is well known that elevated and narrow objects are often
struck by lightning such as tall buildings, church spires and
radio masts. These are also often the focus of corona discharge
although the association between the two is not as strong as
might be supposed. Corona discharge does not necessarily pre-
cede a strike. But it does indicate the presence of a strong electric
field.

There are cases where aircraft have been known to initiate
lightning strikes. Larger craft have a lower electric field threshold for lightning compared with smaller craft and are therefore
more vulnerable. This is believed to result from their higher
capacitance with the negative charge building up during contact
with ice or rain. Lightning and corona discharge have also been
observed during volcanic eruptions but the mechanism of charge
separation is not so well understood.

4.10 LIGHTNING CONDUCTORS AND PREVENTERS

The use of lightning conductors extends back to the mid-18th century but they did not come into general use until later in the 19th century. The idea was first developed by the Czech scientist Prokop Dviisch in 1754. He built a *machina meteorologica* containing over 200 sharp points to help disperse the charge on buildings. For many years controversy raged over their mode of action.[11] There were two schools of thought. Some believed that the conductors acted to reduce the likelihood of a strike by neutralising some of the electric charge built up in the atmosphere. The corona and brush discharges produced by electric storms, it was argued, would gently neutralise the strong electric fields, reducing the chance of a strike. Others thought that the conductor would simply act as a safe and efficient way of conducting the lightning bolt to the ground. The American physicist John Zeleny reviewed the evidence[12] and showed that individual lightning conductors were unlikely to neutralise sufficient current to prevent a strike. He related an incident in Switzerland, when during a thunderstorm the trees of an entire forest glowed with St Elmo's Fire but did not prevent a lighting strike.

Despite the unproven effectiveness of conductors as strike preventers, several manufacturers have produced arrays of barbed wire known as 'eliminators'. But aluminised glass fibres or 'chaff' have sometimes been dropped from aircraft into thunderstorms and appear to help prevent damaging strikes.

Ball lightning also appears to be associated with St Elmo's Fire. The physicist Guy Murchie suggested that ball lightning may be a form of corona discharge advancing along a strong electric field.[13] It differs from St Elmo's Fire as it is not attached to a conductor and has freedom of movement whereas Elmo's is usually attached to a conductor. However, intense radiation can produce an unattached corona in air. A laser beam produces an emission of light caused by the intense electric field at a point of focus in air. This is because electromagnetic radiation is a combination of magnetic and electric fields. When focused the fields can be sufficiently intense to cause ionisation and excitation. The American physicists Vladimir Rakov and Martin Uman have drawn attention to the recently discovered discharges known as sprites occurring high in the atmosphere above thunderstorms. Although poorly understood they appear to be a kind of corona phenomenon featuring nitrogen ionisation. Corona-like filaments have been observed and probably result from fast-moving electrons colliding with air atoms at low atmospheric pressure.

Dry sand blown into the air can cause electric discharges and other, unresolved luminous phenomena.[14] A recent study in the Saharan Desert has shown that the generated electric fields are caused by the particles hitting each other.[15] Here the larger particles acquire a positive charge while the finer dust becomes negative and separation occurs as the finer particles rise higher into the air. It also appears that the presence of the field causes the particles to remain in the air for a longer period of time than they would if uncharged. Since sand consists primarily of quartz grains, piezoelectricity may also be involved (see Chapter 6). A similar process could occur with blown snow and ice and help explain the encounters in the Swiss Alps and Glencoe described above. Some light emission in these cases may also result from triboluminescence.

4.11   OTHER ELECTRIC SPARKS

Small sparks are commonplace but are more often heard than seen. These discharges are frequently the result of static building up on surfaces until the air resistance breaks down allowing the charge to neutralise. The size of the spark depends on the strength of the electric field. Breakdown is influenced by the air pressure, humidity, temperature and the shape of the points of contact. It requires a potential difference of around 30 kV per cm for sparking to occur in ambient air. These potentials are easily reached during friction on non-conducting surfaces such as items of clothing. The discharges can often be seen at night and in some parts of the country were once known as 'fairy sparks'. Sparks are related to the streamers mentioned in connection with St Elmo's fire but differ in producing an illuminated channel of plasma while streamers are only illuminated at their tips. The distinction appears to be somewhat blurred since the crackling sounds heard in some corona discharges are probably due to small sparks rather than streamers. Sparks can also produce impressive Lichtenberg figures as shown in Figure 4.9.

Figure 4.9 Lichtenberg figures appearing on the cockpit windscreen of a jet aircraft flying over New York State. The discharges are probably occurring on the surface of the screen that has been made conducting by precipitation. Reproduced with permission from Ed Kraus.

Sparking from cars is also well known though little reported. It is normally the result of static electricity building up between the occupants and the car seat due to friction. Fuel fires at filling stations are occasionally thought to be due to this, as the electrically-charged occupier touches the fuel dispenser. Women seem to be more at risk. With automatic fillers, drivers often re-enter the car while waiting for the tank to fill, where they may become recharged with static. There are some dramatic images of this occurring on filling station forecourts caught by security cameras. Clothing such as nylon probably increases the risk. When the pump is withdrawn from the tank, a spark may pass. For the same reason, petrol cans carried in cars should first be 'grounded' before they are filled with fuel. In former times certain car models appeared to be particularly prone to static fires, but modern cars are fitted with conducting materials in the tyres to help the charge dissipate. A similar problem can affect aircraft, particularly those built largely of plastic where conducting materials have to be introduced to allow any charges to dissipate.

Martin Van Marum of Holland built a large frictional plate machine to hold electric charges so great that a pointed conductor over 8 metres away showed a corona discharge. John Cuthbertson the English physicist and instrument-maker constructed a machine from Van Marum's designs in 1784 and it can be seen in Teylers Museum, Haarlem (Figure 4.10). The plates consisted of two vertically-set rotatable glass discs 1.65 m in diameter, each of which was excited by four rubbers or

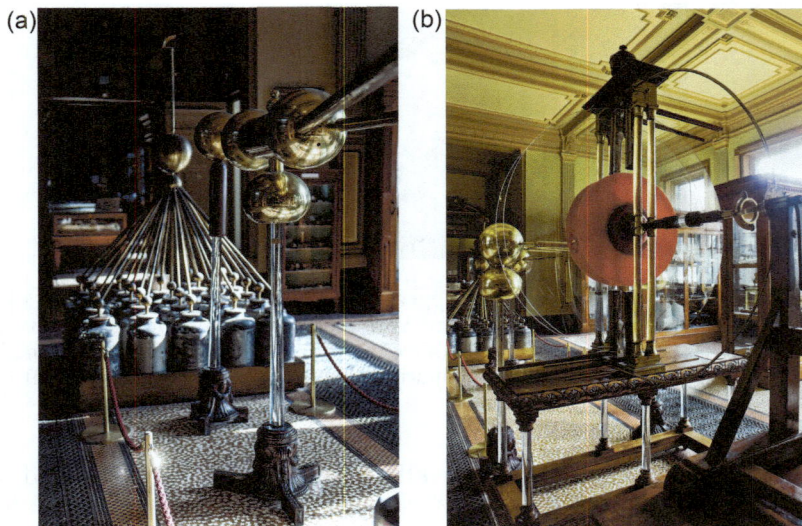

Figure 4.10   Van Marum's electrostatic generator. (a) A bank of Leyden jars used to store electric charge. (b) Large glass discs used to generate high voltages. Courtesy of Teyler's Museum, Holland.

cushions, the principle being that of charge separation – a much enlarged and improved variant of rubbing amber with cloth. The machine could generate around 330 kV. It was attached to the largest battery of Leyden jars ever built and produced a heavy current when discharged.

4.12   GEOMAGNETIC STORMS AND AURORAS

Geomagnetic storms consist of a cloud of plasma consisting mainly of electrons and protons violently ejected from the sun's outer layers. When these plasmas reach the earth, light emission is observed on a far larger scale than that encountered with corona discharges. The astronomer Richard Carrington had sufficient means to build his own observatory at Redhill in Surrey where he carried out extensive measurements on sunspot activity. On September 1, 1859, he noted a pair of bright crescent-like shapes near a large group of sunspots. It was an enormous solar flare and Carrington may well have been the first person to see one. During what became known as the Carrington Event, a huge geomagnetic storm originating on the sun's surface made contact with Earth where sparks leaped from telegraph operator's equipment, some of which started fires and melted wires. Large electric currents were induced in the telegraph cables as magnetic field pulses from the sun reacted with the Earth's field. Auroras were reported as far south as Italy. Within a few minutes of Carrington's observation, the magnetometers stationed at Kew and Greenwich in London were seen to shake violently. This was caused by the first wave of charged particles moving at close to the speed of light reaching the earth. It was followed almost a day later by huge changes in the Earth's magnetic field resulting in damage to wires and cables around the world from a mass of slower-moving particles. In one case, a telegraph cable whose battery had been switched off began communicating messages caused by an induced electric current. This suggested that auroras were connected in some way to electricity. These rare events are now known as coronal mass ejections or CMEs.

In St Elmo's Fire, the flow of charge from the conductor is too small for the magnetic field to have any great influence on its surroundings. But as an ejected cloud of plasma approaches Earth, the accompanying magnetic field is influenced by the earth's own field and modifies its form leading to the erratic behaviour of compass needles together with more serious effects. As the pulsing field cuts across long conductors such as telephone wires and electric cables, an electric current is generated within them. Transformers connected to transmission lines are especially vulnerable as they are easily overloaded. Another CME in 1989 caused the electricity grid to fail in Quebec, damaged transformers and led to a long power cut over most of the province. They can also seriously damage satellites by charging up vulnerable areas causing heavy currents to flow. There is now a satellite early warning system in place to help reduce the most severe effects of these storms.

Auroras are caused by the sun's fast-moving electrons colliding with atoms and molecules high in the atmosphere. Light emission begins more than 100 km above the earth, an observation long known from the work of the Norwegian mathematician Carl Stoermer. He organised observations from a range of sites in Norway using stars as markers and determined aurora height using trigonometry.

Auroras are more noticeable during sunspot activity when the flow of plasma from the sun, the solar wind, is at its greatest. Fast-moving plasma electrons are capable of knocking another electron off an atom or molecule changing it into a positive ion such as N_2^+. Eventually, the free electron will be regained by the ion to form a neutral atom or molecule once more. During the process, a photon of light is often emitted. Other electrons can collide with an atom or molecule without ionisation but by giving up some of its energy, the atom/molecule is converted into an excited state. During de-excitation, a photon is again often emitted.

Fast electrons have energy as a result of their high speed. Those departing from the sun's surface move at about 400 kilometres per second, but may be further speeded up in the Earth's magnetic field. Many of the fast electrons that cause auroras are estimated to have energies of around 1000 electron volts, easily enough to cause the emission of both ultraviolet and visible photons from atoms and molecules.

Aurora light is usually coloured and this indicates that the emitted photons have particular wavelengths and energies (Figure 4.11). The atoms of nitrogen molecules, N_2, are tightly bound together but a fast electron can cause ionisation of the

Figure 4.11   Aurora over Whitby on the Yorkshire coast, autumn 2024. Green and red emissions from oxygen and nitrogen. Reproduced with permission from Andrea Pentecost.

molecule with the emission of a photon of blue light. Nitrogen can also undergo excitation, and photons of red light are then emitted during de-excitation.

These processes help to explain some of the colour of auroral light but oxygen, the second most abundant constituent of the atmosphere is also involved. At a height of 70 km or more above the earth, the atmosphere is extremely thin. In this region, in addition to oxygen molecules (O_2) there are oxygen atoms (O) resulting from fragmentation of the molecules by the solar wind and cosmic radiation. When the oxygen atoms are struck again by fast electrons, they undergo excitation and emit green, yellow or red light according to the amount of energy absorbed during the collision. In these atoms the emission of red light can be delayed by a minute or more since de-excitation involves a 'forbidden' transition. The low gas density above 70 km usually means that the emission of light is faint. Green emission by oxygen is therefore seen more often at lower altitudes. Here the gas density is higher and this has several consequences. The emission tends to be stronger and because the atoms and molecules in the gas are closer together, excited molecules of nitrogen can transfer energy to oxygen atoms causing further emission. This increase in density also means that excited oxygen atoms collide with other gas atoms more often. In the process the extra energy can be passed on to another atom then radiated as heat. Red emission is reduced owing to the time delay of the forbidden transition in a process called collision quenching. Descending further into the atmosphere, the amount of atomic oxygen falls as the energetic electron concentration originating from the solar wind declines. But excitation and ionisation of nitrogen molecules still occur with emission of both red and blue light. The aurora is eventually extinguished as the number of energetic electrons falls to low levels.

The occurrence of auroras around the poles of the Earth is caused by the influence of the Earth's magnetic field. The interaction between the earth's field and the magnetic field associated with the solar wind is complex and auroral patterns are still not fully understood. The earth's field plays a major role in guiding and accelerating the solar electrons towards the earth but it is far from uniform. The earth and sun are rotating at different rates and in different planes. The intensity of the solar wind is always changing and weather patterns on earth can also influence events. The latter appear to be responsible for the recently discovered dune auroras.

FURTHER READING

C. K. Adams, *Nature's Electricity*, Tab Books, Pa., 1987.

J. D. Barry, *Ball Lightning and Bead Lightning*, Plenum, New York & London, 1980.

W. R. Corliss, *Handbook of unusual natural phenomena*, Anchor Press, New York, 1983.

R. A. Ford, *Homemade Lightning. Creative experiments in electricity*, Tab Books, Pa, 1991.

L. B. Loeb, *Electrical coronas. Their basic physical mechanisms*, University of California Press, Berkeley & Los Angeles, 1965.

E. Yaffa and E. Shalom, *The Fourth State of Matter. An Introduction to the Physics of Plasma*, Adam Hilger, Bristol, 1989.

REFERENCES

1. J. E. S. Bowman, Elmos near Glencoe, *Weather*, 1949, **4**, 197–198.
2. E. Whymper, *Scrambles amongst the Alps*, John Murray, London, 1871.
3. R. S. Watson, Notes and queries (electrical activity), *Alp. J.*, 1863, 142–143.
4. C. M. Cade and D. Davis, *The taming of the thunderbolts*, Abelard-Schuman, London, 1969.
5. P. S. Callahan and R. W. Mankin, Insects as unidentified flying objects, *Appl. Opt.*, 1978, **17**, 3355.
6. W. F. Ganong, The Fact Basis of the Fire (or Phantom) Ship of Bay Chaleur, *Bull. Nat. Hist. Soc. N. B.*, 1906, **24**, 419–423.
7. J. M. Jenkins, P. Olson, P. D. McFarland, D. O. Miller and W. H. Brune, Prodigious amounts of hydrogen oxides generated by corona discharge on trees, *J. Geophys. Res.*, 2022, **127**, DOI: 10.1029/2022JD036761.
8. R. Budakian, K. Weninger, R. A. Hiller and S. J. Putterman, Picosecond discharges at a moving meniscus of mercury on glass, *Nature*, 1998, **391**, 266–268.
9. V. A. Rakov and M. A. Uman, *Lightning physics and effects*, Cambridge University Press, 2003.

10. R. F. Orville, A high-speed time-resolved spectroscopic study of the lightning return stroke. Part 2. A qualitative analysis, *J. Atmos. Sci.*, 1968, **25**, 839–851.
11. D. W. Zipse, Lightning protection systems: advantages and disadvantages, *IEEE Trans. Ind. Appl.*, 1994, **30**, 1351–1361.
12. J. Zeleny, Do lightning rods prevent lightning?, *Science*, 1934, **79**, 269–271.
13. G. Murchie, *Song of the Sky*, Riverside Press, California, 1954.
14. H. I. Jensen, Remarkable meteorological phenomena in Australia, *Nature*, 1903, **67**, 344–345.
15. K. Ardon-Dryer, V. Chmielewski and E. C. Brunig, *et al.*, Changes in electric field, aerosol, and wind covariance in different blowing dust days in West Texas, *Aeolian Res.*, 2022, **54**, DOI: 10.1016/j.aeolia.2021.100762.

CHAPTER 5

Radon and Radioactivity

5.1 INTRODUCTION WITH SOME ACCOUNTS

Several phenomena occur in the atmosphere of caves and mines as a faint glow with diffuse edges but do not appear to be electrical in nature. The example below is from David Hodgson. He was an experienced cave researcher in the Yorkshire Dales which contains one of Europe's deepest potholes, Gaping Gill.

For a few days during summer, local caving clubs divert the stream that descends Gaping Gill and set up a winch with a bosun's chair allowing access down the 110 m shaft to the Main Chamber. *Two cavers, fascinated by the huge tunnels and maze of side-passages, were late returning to the Main Chamber. It was midnight and the eerie glow of moonlight faintly illuminated the cavernous space big enough to house St. Pauls Cathedral. Young Fred was the last to leave the chamber using the bosun's chair. Waiting for the chair to descend, he saw a light on the boulder slope at the edge of the chamber. Thinking it was a caver wanting to use the winch he shouted out a warning that it was about finish, but when there was no reply he plucked up his courage and went over to investigate, but as he did so, the light vanished. Believing he had seen a ghost, Fred panicked and ran back to the bosun's chair that had just descended, blew two blasts on the whistle as a signal to*

Luminous Phenomena: A Story of Spontaneous Combustion, Phosphorescence and Other Cold Lights
By Allan Pentecost
© Allan Pentecost 2025
Published by the Royal Society of Chemistry, www.rsc.org

ascend and went up without safety belt or waterproofs in order to get out as soon as possible!

A number of other cave-associated lights have been reported in the UK from time to time. Paul Devereux in his book *Earthlights* mentions the Craven Fairies of Yorkshire appearing as dancing lights in the night sky. They were reputed to live in the caves on Fountains Fell and in the Ribble Valley. He recounted a story by Alex McClennan who described a faint glow in Went Cave above Kettlewell. Here green light was observed to grow more intense as it moved towards him, associated with a rumbling sound. He left the cave in a hurry rather than investigate further.

Lights have also been reported occasionally in mines and they may sometimes be connected with spontaneous combustion as previously noted. An unusual account comes from a visitor to the abandoned Dyliffe lead mine in Wales where the explorer encountered a sudden humming noise from the other side of a pool. Switching off his light a white or blue shape the size of a small man was seen to glow. He quickly retreated to the outside. In *Mineralogia Cornubriensis* published in 1778, William Price noted that the Cornish miners believed that ore veins could sometimes be detected by the appearance of 'fiery coruscations' above them, but no first-hand account was provided. This appears to have been a widespread belief of the miners. They thought it resulted from a flammable exudation from mineral veins. Some of these accounts could probably be dismissed as caver's lights shining up into the sky or afterimages in the eye. The fact that radon can be become concentrated in caves and also emits light is another possibility.

5.2   RADON AS A SOURCE OF LIGHT?

In the case of St Elmo's Fire it has been seen that light emission results from rapidly accelerated electrons striking atoms with the subsequent excitation leading to light emission. Another process leading to light emission is the production of fast atomic particles by radioactive elements. This is termed radioluminescence. The discovery of radioactivity is attributed to the French scientist Henri Becquerel who found that uranium blackened a photographic plate in the absence of light. It was later discovered that some of the atoms of uranium were producing fast atomic particles that struck the plate. They acted upon the plate in a similar manner to light. After much research, radioactivity was found to result in the formation of three types of particles named alpha-, beta- and gamma rays. A large number of isotopes of the elements are now known to be radioactive, but their rate of breakdown varies enormously. Some atoms are so unstable that their lives can be measured in a fraction of a second while others are measured in millions of years. The rate of atom decay is described by its half-life. This is the time taken for half of the atoms to disintegrate. The radioactivity of uranium for example is attributed to its fragmentation into other elements with the emission of high-speed alpha particles. The half-life of uranium-238, the main isotope, is 4.5 billion years, about the same age as the earth. This low rate of decay is made up by its frequent occurrence in the earth's crust.

This chapter investigates the radioactive gas radon. Radon is produced during the breakdown of atoms of uranium, thorium and radium. Radon gas is produced continuously in the crust and seeps into rock crevices so that some of it eventually enters the atmosphere. The gas is highly radioactive. A tube of it would lose half of its atoms within 3.8 days.

The common form of radon, known as radon-222 turns into an atom of polonium and emits an alpha particle in the process. The polonium atoms are even more unstable than radon and have a half-life of just over three minutes. Polonium atoms decay with the emission of another alpha particle plus a beta particle and turns the polonium into yet another radioactive atom – an isotope of lead.

Four more radioactive decays follow so radon decay is seen to be a complex process, with the emission of energetic alpha and beta particles plus some gamma ray photons. Apart from radon, all of the radioactive products of this decay chain are metals. These can be absorbed onto surfaces and may be removed from the atmosphere. With such complex processes going on, it is difficult to predict levels of radioactivity in the atmosphere after radon has been released, but it is known that the level of radioactivity reaches an equilibrium value after a few hours. At this point the concentration of the radioactive products equals that of the radon remaining.

Radon concentration in the air is extremely low and is measured in terms of its radioactivity, a quantity that can be measured fairly easily. The decay of a radon atom, with the emission of an alpha particle is considered as a single radioactive event. Within a given volume of air, one event per second is defined as a becquerel (bq). To put atmospheric radon levels into perspective, radon concentrations in caves and mines are of the order of 2000 bq per cubic metre (equivalent to about 4×10^{-13} grams of the gas) while values in the open air are very much lower. Values within rock crevices could be much higher. In Cresswell Crag Cave, Nottinghamshire levels range from 27–7800 bq m^{-3} with seasonal variations. Higher values tend to occur deeper in caves, although shallow Scoska Cave in the Yorkshire Dales has recorded levels in excess of 12 000 bq m^{-3}. A study of south-west England mines and caves gave averages of 10 000 and a maximum of 800 000 bq m^{-3}. Levels in winter tend to be lower owing to the ingress of fresh air into caves. In contrast, during fine summer weather when high atmospheric pressure prevails, radon concentrations may increase, particularly under atmospheric inversions.

There has been recent interest on the possibility of using soil radon to predict earthquakes. Radon levels behave anomalously in soils and ground waters during seismic activity perhaps due to the opening and closing of cracks in the strained rock. It has also been noted that radon levels are generally higher over faults. Thus there could be a connection between radon, light emission and seismic activity.

To gain some idea of the likelihood of light emission by radon in caves, three of us made a visit to Scoska Cave in Yorkshire (see Box 5.1). A single alpha particle can produce about 100 photons in air, but is there enough radon emission to produce visible light?

5.3   RADIOLUMINESCENCE IN THE OPEN AIR

During the Chernobyl Incident, workers reported beams of blue light rising into the sky above the destroyed reactor resulting from intense gamma ray emission from the highly radioactive fission products. Blue glows lasting several seconds have also been observed after nuclear explosions in the atmosphere. Radioactive particles impacting nitrogen molecules in the air leads to a short period of excitation. During de-excitation photon

---

BOX 5.1   SCOSKA CAVE

On a moonless January night we set up a Nikon film camera on a tripod 40 metres within the cave pointing away from the entrance toward a small table upon which stood a shadow marker and a piece of fluor-spar (Figure 5.1). The camera was equipped with an F1.4 lens and an exposure of 90 minutes was made with ASA 400 film. To avoid breathing in too much radioactive air, a guide rope was fixed down the cave so that the film could be exposed without light contamination.

Figure 5.1   (a) Looking into the depths of Skoska Cave, Yorkshire showing the set-up for radon photography. (b) Shows shadow marker and fluorspar on the table.

A shadow marker was used to make sure no extraneous light entered the chosen area. The fluor-spar was used to check for ultraviolet radiation as the mineral fluoresces in its presence. Our results were negative. We did not expect to detect light from the radon once it had dispersed into the cave atmosphere but hoped to find it in small fractures in the cave wall where much higher values should occur. Radon levels in the cave at this time were about 2500 bq per cubic metre – we would have liked more. Interestingly, the cave was host to a large number of tissue moths (*Triphosa*) hibernating over winter and apparently unaffected by the radiation.

emission occurs in the visible region, mostly in the blue part of the spectrum. Excited oxygen is also produced but here the ultimate product of de-excitation tends to be ozone (O_3) and light emission is less significant.

A more familiar example of radioluminescence can be found in the luminous dials of clocks and watches. Luminous paints were made from a mixture of a radium compound and a luminescent solid such as zinc sulphide doped with copper. The alpha particles would collide with the copper atoms leading to excitation and emission of a bluish-green light. Alpha rays have a range of about 4 cm in air and a much shorter range in solids so the wearer of a wrist watch for example would not be seriously affected by them. Unfortunately, radium also produces much more penetrating gamma rays. Once this health hazard was recognised, other less damaging substances were used. Some of the earliest breakthroughs in particle physics were achieved by examining emissions from a zinc sulphide screen using a radioactive source. Modern watches have their luminous components made of small glass tubes containing the radioactive gas tritium, an isotope of hydrogen that produces β-particles. The light emitter is painted on the inside of the tube. Tritium is a weak emitter so health hazards are minimised. Radioluminescence can also be seen through an astronomical telescope. Fast atomic particles from supernovae passing through vast clouds of gas in the galaxy become lit up in a spectacular fashion.

FURTHER READING

M. Kerrigan, *Chernobyl*, Amber Books, London, 2022.

R. Lauf, *Introduction to Radioactive Minerals*, Schiffer, Lancaster, Penn, 2008.

R. C. Malley, *Radioactivity. A history of a mysterious science*, Oxford University Press, 2011.

A. M. Stacks, *Radon. Geology, Environmental Impact and Toxicity Concerns*, Nova Science, New York, 2015.

CHAPTER 6

Luminous Minerals, Stones and Powders

6.1 INTRODUCTION

This chapter includes examples of phenomena relating to rocks and minerals, namely their fluorescence and phosphorescence resulting from photoluminescence, as described in Chapter 1. In addition the emission of light through fracturing and other pressure effects is included – examples of triboluminescence, electroluminescence and sonoluminescence plus the emission resulting from the heating of minerals known as thermo-luminescence. They include some of the first phenomena of light emission to have been examined experimentally.

It was only possible to appreciate the fluorescence of minerals once a way could be found to produce ultraviolet radiation in-dependently of the visible radiation. This began when glass prisms were used to split the rays of light from the sun. By il-luminating certain objects beyond the visible spectrum some of these invisible radiations sometimes produced a striking fluo-rescence. We begin by examining some examples of phosphor-escence since these discoveries preceded those of fluorescence.

Luminous Phenomena: A Story of Spontaneous Combustion, Phosphorescence and Other Cold Lights
By Allan Pentecost
© Allan Pentecost 2025
Published by the Royal Society of Chemistry, www.rsc.org

6.2   PHOSPHORESCENT MINERALS AND STONES

There are many intriguing reports of phosphorescence in natural stones. These relate mainly to gem stones, long recognised as objects of beauty with those shining at night receiving particular attention. The Temple of Melqart–Heracles in Tyre dates from around the 10th century BCE and was dedicated to the Phoenician patron deity and god Melqart. It was visited by the writer Herodotus in the 5th century BCE and described as having two of its columns adorned with precious materials, one of gold and the other of emeralds. The latter was described as glowing in the dark and was therefore phosphorescent. In his historical survey of luminous gems, S. H. Ball[1] suggested the columns may have been equipped with lamps placed behind a coloured glass instead. There are many other stories relating to luminous stones. One of the Jewish Talmudic legends states that Noah's Ark was lit at night by a luminous stone to assist its passage.

Rubies have also been noted as phosphorescent. These were known as *anthrax* (glowing coal) to the Greeks and *carbunculi* to the Romans, but other red-coloured gems were probably included as well. One of the most famous was a large stone owned by a king of Ceylon but is now lost. The story began in the 6th century CE when it was described as a stone as big as pine cone. At some period it was thought to have been placed in a temple near Anuradhapura in Sri Lanka (formerly Ceylon) and early travellers reported it as having the appearance of a glowing fire. Rubies are coloured red by traces of chromium and fluoresce in ultraviolet light. Since sunlight contains some UV this may be the origin of the story although some rubies are also known to phosphoresce. Rubies are crystals of corundum (aluminium oxide), and the chromium ion sits awkwardly in the host's crystal lattice, replacing a smaller aluminium ion. This distorts the lattice leading to the absorption of violet and yellow-green light giving the mineral its rich red colouration. Ruby lasers are based on this phenomenon. Historically, some 'rubies' were actually spinels which they can closely resemble. 'Ruby' spinel is sometimes found associated with the true ruby in gem deposits. It is an aluminium silicate containing traces of chromium but spinels do not appear to phosphoresce. Chromium, one of the transition elements, also plays a role in the unusual colouration of a form of chrysoberyl known as alexandrite. Clear crystals of good quality alexandrite appear greenish in daylight but red in artificial (incandescent) light owing to strong light absorption in the yellow region of the spectrum by chromium ions. Some artificially-produced gemstones doped with cobalt show a similar effect along with several chemical compounds such as anhydrous chromium sulphate and thulium(II) salts.

Garnets are also known to be phosphorescent. At Visby on the island of Gotland in the Baltic Sea, two large garnets were placed in the centres of rose windows in the church of St Nicholas and were said to shine brightly at night, guiding mariners to the port. They did not survive for long. The town was attacked by Valdemar IV in 1361 and the jewels stolen, only to be lost on the Karlsö Islands soon after in a shipwreck where presumably they still remain.

There are early stories about miners who were able to locate gems by their phosphorescence. S. H. Ball pointed to an old Banyan legend where the first diamond mine was discovered owing to the brightness of its stones. In the *Physiologus*, an ancient text, it is also noted that the diamond is only found at night. One remarkable diamond deserves special mention. The Hope Diamond was purchased in India some time in the 17th century by the merchant Jean-Baptiste Tavernier. It weighed 45 carats and had an unusual blue-grey colour. It was later discovered that the stone showed a strong red phosphorescence after being irradiated with UV. This was probably connected with its colour, caused by traces of boron. Like many large gems, the stone has had a turbulent history. It was sold to the French Royal Family but stolen during the French Revolution eventually making its way to London. At some stage it was re-cut and ended up with the British Royal Family and then a London banker, Thomas Hope. The Hope family was forced to sell to cover debts and it was acquired by a New York dealer. After several private owners in the United States it ended up in the National Museum of Natural History, Washington DC, where it may be seen today.

Fluorite (calcium fluoride) is a common mineral well known for producing large and attractive cubic crystals (Figure 6.1). Most specimens show a bright violet fluorescence under ultraviolet radiation attributed to traces of lanthanoid metals in crystal defects. A much less common yellow fluorescence has

Figure 6.1   Green fluorite crystals from Weardale, Durham. Natural History
Museum mineral collection. (a) Large green crystal with an edge
of 60 mm in length. BM 31354. (b) A dark green twin, purple in
reflected light. 35 mm wide. BM 88934.

been attributed to the element dysprosium. David Brewster
(1781–1868) noted that light emission in fluorite took place
along crystal planes and suggested it was due to some substance taken up during crystallisation. The Irish physicist
George Stokes (1819–1903) after a series of well-designed experiments, demonstrated a connection between light emission
and ultraviolet radiation and coined the term fluorescence after
this mineral. Phosphorescence in fluorite is less often reported
and does not appear to be common although crystals found in the
Levant are said to occasionally show it. On Cyprus, the tomb of
King Hermias of Atarnus (*ca.* 341 BCE), was said to have been
adorned with a marble lion which at night shone with green eyes
thought to have been made of chlorophane, a form of fluorite, but
nothing now remains. More recently, the mineralogist Gustavus
Rose observed that chlorophane gravels shone brightly at night on
the Irtysh River in the Altai region of Russia whilst on a geological
expedition.[2]

Chlorophane sometimes luminesces faintly in the hand and
possesses properties of thermo-, tribo-, and fluorescence
(Figure 6.2a and b). It might have been the phosphorescent
'emerald' referred to in the Temple of Melqart–Hercules mentioned above. H. E. Millson, father and son, undertook some
studies on the long-term phosphorescence of minerals.[3] They
illuminated a sample of fluorite from Trumbull, Connnecticut
and a sample of calcite from Texas with a burst of UV from a
mercury vapour lamp and detected their phosphorescence using

Figure 6.2 Chlorophane crystals from Aldan, Siberia. One crystal shows a weak phosphorescence under illumination. Main crystal edge of 7 mm wide. Natural History Museum mineral collection. No. 89004.

'phosphography'. They placed their minerals on photographic film in the dark for long periods of time. Both minerals were found to phosphoresce 18 years after irradiation. Phosphorescence in the fluorite was visible to the eye more than 4 years after the irradiation. They discounted radioactivity as the source of the light emission as they found the phosphorescence to be temperature sensitive. Although fluorite is a fairly common mineral often associated with lead and zinc ores, reports of its phosphorescence in mines following illumination do not appear to have materialised. They may be due to the low levels of UV in most forms of artificial light. There are other stories of night-shining minerals. Parthenius (*ca.* 10 BCE) stated that a stone known as the 'Aster' was to be found in the Sakarya River of Turkey that flamed in the dark. In 632 CE it was found that jade deposits in the beds of the Hotan River system of Turkestan (Tarim Basin, China) shone at night, allowing divers to recover them.

A small number of other minerals are known to phosphoresce. Tugtupite and hackmanite, forms of sodalite are among them, the latter also being known for its tenebrescence. Others include some forms of borax, celestine and lorenzenite, the latter sometimes showing a brilliant green glow when excited by UV radiation.

There are several reports of water ice phosphorescing in the form of snow but little research has been done. Some of these may relate to electro- or triboluminescence. In the *Alpine Journal*

E. Holland described bright phosphorescent sparks originating from freshly fallen snow in the Austrian Alps as it was disturbed by walking.[4] There was no other source of light in the sky. Marcel Minnaert mentioned an old legend describing ice fields giving out a feeble light at night after having been exposed to bright sunlight.[5] Cold snow has also been seen to glow when brought into a darkened room. In the winter of 2021, phosphorescent snow was observed on the shore of the White Sea in Russia but closer examination revealed the presence of the bioluminescent copepod *Metridia* which had become trapped within it. Other luminescent organisms could also have been involved (Chapter 7).

6.3   LUMINOUS POWDERS

In a Chinese text of the Song Dynasty (960–1279 CE) there is a reference to a book, now lost, suggesting that phosphors were known in China as early as the Han Wu Di Period, 140–88 BCE. The same text describes the acquisition of a painting by Xu Zhi-e. The picture showed a cow grazing beside its shed but this was no ordinary painting. When viewed in the dark, the cow was seen asleep inside the shed and was made possible by the application of phosphorescent paint. The painting was later passed to Zhan Nin who related that in the China Sea, nacre, the pearly deposit found in the shells of some molluscs, sometimes contained a liquid that could be turned into a luminous ink. This suggests some form of bioluminescence, although nacre itself does not appear to behave in this way. More likely, the molluscs had been feeding on bioluminescent plankton (Chapter 7). If this was the source of the phosphorescent paint then it would probably have been rapidly denatured once it had dried. Another possibility is that the shells were heated to a high temperature, producing some calcium sulphide which can occasionally show a red phosphorescence. This would be admirably suitable for painting a sleeping cow, although the paint would again become degraded over time. A possible scenario for this picture is shown in Figure 6.3.

Monte Padurno close to Bologna, Italy, is a popular site for visitors. Hand-sized lumps of the mineral barite (barium sulphate) occur in clay soils and must have attracted attention

Figure 6.3 Reconstruction of a Chinese painting of a cow and its shed. Viewed in a lighted room it is seen grazing outside its shed during the day. When the room is darkened it can be seen asleep on the hay.

from early times. The material consists of concretions that appear to have grown around nuclei present in the soil (Figure 6.4). A good number were collected by the alchemist Vincenzo Casciarolo who undertook a series of experiments to understand their properties in a search for the mythical Philosophers Stone.

The high density of barite would be certain to draw attention, and alchemists might have assumed it contained some form of gold, an exceptionally heavy metal. A good account of Cascariolo's discoveries was published by Fortunius Licetus in 1640.[6] In it he describes how the stone was first reduced to a powder and made into cakes with either water or egg white. The cakes were then fired at a high temperature for several hours. The resulting powder, once exposed to light, glowed red in the dark, no doubt much to the amazement of its discoverer. The astronomer Galileo Galilei

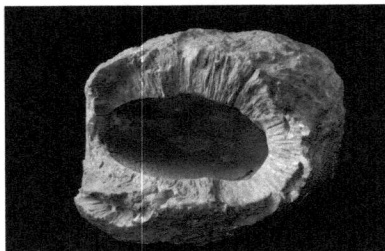

Figure 6.4   A sample of the Bologna Stone broken open to reveal a central void probably occupied by a foreign stone that acted as a nucleus. Radiating crystals of barium sulphate have grown upon its surface. Stone *ca.* 20 cm wide. Courtesy of the Alma Mater Studiorum of Bologna, Luigi Bombicci Museum. Image by Francesca Bargossi.

was shown the calcined material and concluded that fire and light became trapped within it and was slowly released like the absorption of water by a sponge. In this, Galileo's interpretation is not far removed from what is currently understood. Licetus was convinced that the light of the moon was also the result of such phosphorescence and disputed Galileo's interpretation that it was light reflected from the sun. A more detailed account of its history is provided by Roda.[7] The chemical constitution of barite was slow to realise but by 1800 it was found to contain sulphur and a new heavy element now called barium. Through the work of Vauquelin and others it soon became apparent that barite was barium sulphate, $BaSO_4$. Calcination with a reducing agent such as charcoal removed the oxygen from the sulphate resulting in barium sulphide and this was the compound which was phosphorescent, although pure samples did not show the property. This led to the suggestion that some impurity associated with the sulphide was responsible. Barium sulphide is also unstable in moist air and the phosphorescence properties of the powdered stone were found to decline over time as the sulphide gradually reverted back to sulphate. It was therefore apparent that both the sulphide and some other unknown substance was involved.

Barium sulphide is not known to occur as a mineral but calcium, another alkaline earth element, forms a mineral sulphide and the impure compound has been found associated with coal mine fires in Poland as a result of pyrometamorphism. The mineral is called oldhamite. It is also known from meteorites but there appear to be no recorded instances of its phosphorescence. In the 17th century, Christian Balduin dissolved calcium carbonate in nitric acid then dried the residue and heated it to a high temperature. It yielded a reddish phosphorescence and he termed it *phosphorus hermeticus* since it would only glow if preserved in a sealed jar.[8] In or around 1693, Wilhelm Homberg heated a mixture of ammonium chloride with quicklime, (calcium oxide) and obtained a phosphorescent material, presumably calcium chloride although this compound is not noted for its phosphorescence but it has been reported as triboluminescent.[9] His material may have been contaminated with sulphur or one of its compounds. The material was later termed 'Homberg's phosphorus', not to be confused with *Homberg's pyrophorus*, a spontaneously combusting mixture obtained from excrement (see Chapter 2). In a similar fashion, John Canton calcined oyster shells with sulphur and obtained '*Canton's phosphorus*', again a calcium sulphide.[10] None of these materials appears to have resulted in commercial success, probably owing to their susceptibility to breakdown by moisture. It is also of interest to note that the element sulphur itself emits light if gently warmed although few people appear to have reported it. It is known that some oysters when ground to a paste, 'phosphoresce' of their own accord, probably due to the presence of luminous bacteria. Other calcium compounds are known to phosphoresce. Slaked lime or calcium hydroxide has been used for millennia as a pigment and was once used to paint the towers of English churches, improving their visibility. There are suggestions that the material is weakly phosphorescent adding to their increased brightness during both day and night although this does not appear to have been substantiated.

Magnesium and strontium are also alkaline earth elements and it was not long before their sulphides were more fully investigated. Impure strontium sulphide is sometimes phosphorescent. Magnesium sulphide is difficult to prepare though it was later shown to phosphoresce. Once the phosphorescence of the alkaline earth sulphides became established, chemists began working on them. As long ago as the 18th century it was recognised that blue light was more effective in imparting phosphorescence than red and by the early 20th century, P. Lenard and others thought some form of electronic excitation was responsible. The pure alkaline earth sulphides were confirmed as being only weakly or not at all phosphorescent and a thorough search was made to uncover the impurities responsible. By deliberately adding small amounts of other materials it was discovered that a wide range of metals and their salts were effective when added to these sulphides, including copper, lead, thallium, tungsten and vanadium. The sulphides themselves are now regarded as 'host' molecules' playing a minor role in phosphorescence. In order to obtain good emission, the host needs to be 'doped' with one or more 'activators' often at low concentration, of the order of a few parts per million. In addition, 'co-activators' which appear to have a catalytic action, may also be required. The concentration of the activators can be critical and this explains why naturally phosphorescent minerals are rare. Intensive research has subsequently been undertaken on phosphors and has shown that in some cases an exacting set of circumstances is required for success. It has even been reported that two chemists, working with exactly the same recipe and using the same conditions sometimes fail to achieve similar results. Most activators are now known to be metal ions with a valency that differs from the host metal ion(s).

Zinc sulphide also phosphoresces. Theodore Sidot in 1866 produced small crystals of the sulphide by sublimation and noted their phosphorescence, probably as a result of their contamination with a trace of copper. It was known for a time as Sidot's blende. The common mineral sphalerite consists of zinc sulphide but natural samples do not appear to phosphoresce although according to the American mineralogist James Dana some specimens do so if scratched. Natural samples of zinc sulphide (sphalerite) are usually contaminated with iron and this may act as a quenching agent or 'sink' for phosphorescence although the mineral is frequently fluorescent under UV. The rarity of natural phosphors is probably exacerbated by contamination from minerals containing substances that quench the emitted light with further losses resulting from self-absorption – few minerals are transparent.

Zinc sulphide is one of the best known 'phosphors' and was formerly used in applications such as cathode ray tubes and X-ray screens. In the early 20th century, Ernest Rutherford and Hans Geiger developed the zinc sulphide scintillation screen used to count atomic particles. They played an important role in early nuclear physics. In most of these cases the emissions were caused by cathodoluminescence and electroluminescence. The compound when phosphorescent usually emits a green light but it is short-lived by human standards, lasting just a few minutes before fading away. Large screens of zinc sulphide are sometimes employed for entertainment as 'shadow walls' where a person stands in from of a screen which is irradiated with UV. An 'after image' is created which persists for a few minutes after the person has moved away. By adding traces of gold and indium the phosphorescence may be extended to several hours. Other activators give strong red (manganese) or blue (silver) emissions.

It is likely that several different processes are involved in zinc sulphide phosphorescence. Activators such as gold have larger atoms than zinc and are more likely to occupy 'interstitial' sites while smaller activator atoms are likely to substitute for zinc atoms in the crystal framework (Figure 6.5). Even high purity zinc sulphide shows a weak phosphorescence and in this case it is thought that a few additional and isolated zinc atoms in the crystals are responsible for a process described as 'self-activation'. Most phosphors are difficult to prepare and require an electric furnace to provide a sufficiently high temperature for the activators to disperse efficiently in the host. There are exceptions and a group of French chemists have described a bright strontium aluminate phosphor that can be made in a microwave oven in the school laboratory.[11] This phosphor has been used recently

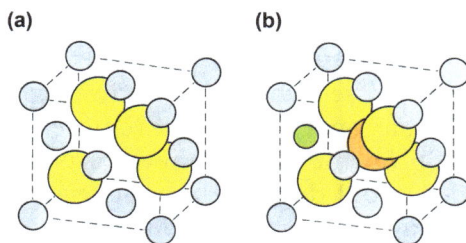

Figure 6.5 The unit cell of cubic zinc sulphide, ZnS (a) The unit cell with 12 zinc atoms (grey) and four sulphur atoms (yellow). (b) A unit cell containing an atom substituting for zinc (green) and an atom in an interstitial position within the crystal lattice (orange). Atom/ion sizes are approximate. Bar 5×10^{-10} m.

to make 'glow in the dark' plants by feeding them with a mixture of the aluminate and traces of a lanthanoid element such as europium.

If host compounds play but a small part in phosphorescence it is reasonable to ask why so few of them are known. In the case of zinc sulphide this may be related to its easy fusibility and the high solubility of some of the activators such as the lanthanoids, allowing the activators to disperse efficiently within them but there are other factors. The relative size and electric charge of the host and activator ion are known to be important. Many other hosts are in fact known and include Group II oxides, sulphates and halides. Detailed models of electron orbitals and their interactions indicate complex relationships between the host, activator and co-activator. Synthesis of phosphorescent material still requires a certain amount of guesswork. It is also worth noting that a number of phosphorescent minerals are also fluorescent and some of the sulphides are candoluminescent.

6.4 TENEBRESCENT MINERALS

Tenebrescence is the result of changes in light absorption rather than emission. It is known in a small number of minerals and chemical compounds. Well known examples include the chlorides of potassium and silver. Potassium chloride crystals are normally colourless and transparent, but when subjected to X-rays or electron beams they become a dark magenta colour which can be reversed upon heating. Silver chloride, also colourless, becomes dark upon exposure to ultraviolet radiation if it is used as a dopant within the glass. Use of this is made in photochromic sunglasses which darken in bright sunlight. The process is reversible once the UV is removed, providing convenient protection for the eyes. However it is still temperature-dependent and on cold days the rate of reversion falls. These phenomena are known to result from crystal defects where a 'hole' with a missing anion (a chloride anion in the above examples) is filled with one or more unpaired electrons. The resulting electronic structure absorbs light in particular regions, colouring the crystal. The holes are known as Farbe centres and have been well studied for their technological applications.

Tenebrescence was first seen in minerals. In 1830, T. J. Pearsall observed that fluorite crystals were discoloured by violet light. Much later, hackmanite obtained from granite and also a form of sodalite was found to change to a purple-red colour on exposure to UV, being reversed once the radiation was removed.[12] It has also been reported for kunzite, a form of spodumene and some scapolites and tugtupites. The deep blue colour of the fluorite known as 'Blue John' is suspected of resulting from a form of tenebrescence. In this case, small inclusions of uranium minerals are thought to be responsible for the irradiation and subsequent colouration. Zincite (zinc oxide), if in a pure state, when heated changes from white to yellow and this is also considered to be a form of tenebrescence. It is also shown by several other metallic oxides such as those of cadmium and a few of the many oxides of caesium, antimony, germanium and titanium.

These substances have been termed scotophors to contrast them with the light-giving phosphors. The phenomenon is also observed in several other chemical compounds, one of the more interesting being the pigment Prussian blue much used by watercolourists, some of whom have commented on its tenebrescence qualities. They are all believed to possess Farbe centres though some have yet to be studied in detail.

6.5 TRIBOLUMINESCENCE AND EARTHQUAKE LIGHTS

Triboluminescence has been observed in rocks for many years. Ceremonial rattles of the indigenous peoples of Colorado were filled with quartz pebbles and emitted light when shaken. In the ruins of Pecos pueblo in New Mexico, a 'lightning stone' was discovered consisting of a quartz cylinder that was made to revolve in a semicircular groove, emitting light. The ease with which triboluminescence, perhaps associated with piezoluminescence, occurs in pebbles of quartz suggests that it should be visible on a dark sea shore with waves crashing on a pebbly beach.

Ice, when removed from a freezer often glows as the crystals warm. During melting, the crystals become stressed and often fracture. Olaf Wässtrom reported the emission of light from broken ice and triboluminescence has also been reported on the bows of icebreakers.[13] This may also be the cause of the 'phosphorescence' observed during snow- and sandstorms as a result of attrition. In some cases however the emission of light has another origin. Koch in 1882 observed that a ship breaking through ice in Hardanger Fjord, Norway, gave flashes of light. When some of the ice was collected and melted, the water remained luminous indicating bioluminescence.

Surprisingly there appear to be no accounts of triboluminescence in mines. Deep mines are occasionally subjected to high stresses leading to collapses. The local fracturing and sliding of material should result in emission of light from a range of processes, notably incandescence, tribo- and piezoluminescence.

Use has been made of the phenomenon in the monitoring of cracks in concrete bridges. A luminescent sensor sheet is placed over the area where a crack is suspected and covered with an opaque sheet to prevent daylight from entering. Sheets are removed and replaced from time to time. As the crack develops it can be visualised through its triboluminescence to reveal how it has progressed. Triboluminescence may also play a role in the more spectacular displays of light seen in earthquakes as described below.

6.5.1 Earthquake Lights

Earthquakes – a series of shock waves generated within the earth are sometimes associated with the emission of light and other radiations. Three classes of earthquake waves are generated; P or 'push and pull' waves are the first to be detected during a quake since they travel at the highest speed, typically around 8 km per second. They are equivalent to sound waves in air. The next to arrive are usually the S or 'shear' waves but they cannot travel through a fluid so are not always detected. Finally there are the long-period or L waves. These are the most destructive and cause the ground to rotate along elliptical paths and also from side to side. These waves die out more rapidly than the P or S waves but can be intense in the region surrounding the source of the shock.

Since the advent of digital cameras and mobile phones there have been many sightings of earthquake lights from around the world although their origin remains a mystery. They add weight to the veracity of at least some of the earlier reports of lights such as the Japanese document of CE 869, the *Nihon Sandai Jitsuroku* regarding the Sanriku earthquake. Lights have been seen before, during and after large quakes but mainly during the events. Their intensity varies and in some cases such as those of the Japanese Idu earthquake they have been recorded from a distance of 70 miles (112 km) or more. The individual illumination events last less than a second to several minutes and most accounts indicate a bluish colour.

In Britain there are at least three published records of the phenomenon.[14,15] The most significant of these relates to the Hereford Earthquake of 1896 which occurred during the night of 17th December (see Box 6.1). The second report comes from the Channel Islands in February, 1927. The epicentre was under the sea to the east of Jersey. It was a fairly significant event as it was felt as far east as London and came about 6 months after a similar but stronger quake near the islands. Although no details have emerged, there were reports of something resembling a meteor in the sky at the same time. The North Sea is well-known for its earthquakes and an exceptionally large one (for Britain) occurred near the Dogger Bank in June 1931. This was witnessed

---

BOX 6.1   THE HEREFORD EARTHQUAKE

On 17th December 1896 at 5.32 am an earthquake originating close to Hereford was felt over most of England and Wales. Although a few seismometers were operating at the time, they were not sufficiently accurate to provide details of the event. The seismologist Charles Davison stepped onto the stage and undertook the most detailed study of an English earthquake up to that time.[16] Davison prepared a questionnaire and sent over 400 letters to newspaper editors throughout the British Isles. He also made local enquiries and contacted people whom he considered both reliable, or belonging to a profession of people who were likely to have been active in the early hours. These included railway signalmen and members of the clergy. As a result he received 2902 accounts of the earthquake. These were then analysed systematically allowing him to map several of the quake's characteristics. He was able to map the earthquake's isoseismic lines (lines going through points of equal intensity) and provide details of its duration, progress of the earthquake waves and the production of sound. Its magnitude has been estimated as 5.2.

In an appendix Davison drew attention to a meteor that was mentioned in some of the accounts. The positions of the observations were described for eight of these. They are shown in Figure 6.6 as red circles. Unfortunately he omitted the remaining accounts probably owing to the fact that the two phenomena were not thought to be closely connected. However, Davison analysed all of these accounts and their descriptions of the meteor's position in the sky. He estimated the most likely track of the object. This is shown by the red dashed lines in Figure 6.6.

Several accounts suggested that the meteor first appeared in the south and then proceeded north. Close to the earthquake epicentre near Hereford (marked by a black dot on the map), the appearance of the meteor coincided with the earthquake. To the south and north, accounts indicated that it appeared just before the quake was felt. The evidence points to an earthquake light occurring in the vicinity of the epicentre coinciding with the shock. Descriptions of the meteor varied, but some observers likened it to balls of fire traversing the sky.

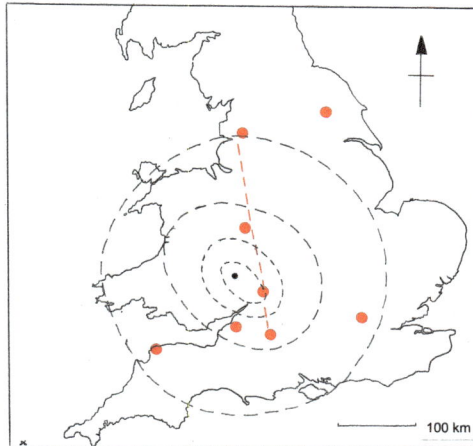

Figure 6.6   The Hereford Earthquake. Isoseismic lines and trace of the 'meteor' (broken red line).

throughout the country, including the east of Ireland and caused considerable damage in eastern England. Again there were reports of lights in the sky at the time of the quake but no details were forthcoming. These lights were no doubt reflections of ground emissions from overlying clouds. The magnitudes of these earthquakes were in the range 5.2 to 6.1. It is among these magnitudes and above that most reports of light emission occur.

At least four explanations have been proposed for earthquake lights: emission from quartz as a result of pressure resulting in piezoluminescence; ionisation of oxygen in minerals resulting in surface charging and subsequent emission; disruption of the ionosphere caused by surface charging and finally, triboluminescence. These topics are considered in turn.

6.5.1.1 Piezoelectricity. Robert Boyle discovered that a diamond subject to intense local pressure can emit light. Later, it was found that a wide range of minerals, including rocks, can become charged with electricity if compressed. This was first demonstrated in crystals of quartz by the Curies and became known as piezoelectricity. Charging occurs because in many minerals, an increase in pressure leads to small changes in the positions of the atoms in the crystal lattice. For some crystals this can lead to an electric charge on opposing faces of the crystal which is lost when the pressure is removed. With quartz a significant voltage can develop which under controlled conditions can result in a spark. Conversely, application of an electric potential across particular crystal faces will alter the dimensions of the crystal.

Piezoelectric materials have been well studied and are found in a wide range of manufactured items such as inkjet printers, sonar detectors, transducers and hand-operated fire lighters. Since quartz is a particularly effective piezoelectric material and is a common mineral in the earth's crust, pressure waves generated by earthquakes would be expected to generate electricity on the surface of the crystals. If the energy of the resulting electrons can be changed into light then earthquake lights may be the result. In a sandstone for example, where quartz is the main constituent, one can imagine sparking occurring across the air gaps between individual grains. For this to occur the rock would need to be dry otherwise any excess charge would be carried away by the conducting water film. Evidently light could only be detected where sandstones occur at the earth's surface since they are opaque. These rocks are rarely of pure quartz and normally contain a substantial quantity of dark, light-absorbing minerals. Although the writer has not found reports of light emission from compressed sandstones or quartzites, friction has been found to cause light emission in sandstones presumably due to triboluminescence, perhaps accompanied by some piezoluminescence. Another possibility is corona discharge on rock surfaces resulting from a strong piezometrically generated electric field. A brief surface charge resulting from an earthquake wave could attract environmental electrons in the air above. These would be accelerated by the electric field and the electrons would collide with air molecules leading to excitation and light emission. This is a form of electroluminescence and calculations by S. Takaki and M. Ikaya indicate that the emitted light should be visible.[17]

Light emission has also been reported during sandblasting operations and in dust storms and a piezoelectric origin has been suggested.

6.5.1.2 Oxygen Ionisation. A group of geologists headed by R. Thériault suggest that earthquake waves cause fractures in rock leading to the breaking of peroxy bonds (*e.g.* $O_3Si–O–O–SiO_3$) with production of oxygen anions (O^-) as the rock becomes semi-conducting.[18]

This could again provide surface charges appearing at the earth's surface, particularly over topographic highs and in the vicinity of vertical or near-vertical faults and mafic dykes. Their research indicated that the majority of quake light sightings are associated with these types of structures, the Ebingen, Pisco Peru and Sageunay earthquakes being cited as examples. The strong electric field may then result in local light emission as a corona discharge.

6.5.1.3 Disruption of the Ionosphere. A study by Friedemann Freund and co-workers suggested that when rocks are stressed they could change from being electrical insulators to semi-conductors causing the release of 'dormant' positive charges.[19] As insulators, the charges would show little movement, but an applied stress may cause them to migrate producing an electric current and magnetic field. If on a suitable scale, the field could influence the ionosphere in the upper atmosphere. Changes in the ionosphere can be detected on Earth as they alter short wave radio signals. Such differences were observed just before the large 1961 Chilean and 1964 Alaska earthquakes. Changes to the ionosphere could also result in aurora-like phenomena. Several accounts of earthquake lights mention their aurora-like form.

6.5.1.4 Triboluminescence.   It has already been noted that the fracturing of crystals sometimes results in light emission so large-scale fracturing of rocks is likely to do the same. Granite for example exhibits light emission when it is crushed in the dark although piezoluminescence may also be involved.

Powerful L waves are known to cause the opening and closing of fissures in the ground and these are likely to lead to mineral fracturing. One might expect some form of general light emission as a result. Experimental studies by Troy Shinbrot using mineral grains have revealed that the opening and closing of fractured mineral grains is accompanied by a positive, rapidly followed by a negative voltage spike. If these processes occur during the passage of an earthquake wave, large potentials might result leading to light emission in the atmosphere.

6.5.1.5 Other Explanations.   The New Madrid earthquake was one example where lights were seen, some being convinced that a volcanic eruption had started.[20] This was a large earthquake that shook Missouri and Tennessee in 1811 and changed the topography of thousands of square miles of land. It resulted in the appearance of Reelfoot Lake, 20 square miles (52 km^2) in area. The earthquake raised dark clouds of dust and there is a possibility that electric discharges within the cloud were occurring.

Several observations report lights being seen over the sea during earthquakes as for example during the large Peru quake of 2007 and the North Sea quake of 1931. If the quake occurred under deep water, say >300 m it is unlikely that much visible light would reach the sea surface. Gases trapped close to the sea floor might also be released during an earthquake. There is an intriguing account of such an occurrence off the coast of England during the 1886 Dartmouth quake, when the nearby sea was said to be 'covered in foam'. Since a considerable time is likely to elapse between the shocks and appearance of gas, it is unlikely to influence any emission of light.

There are two other possibilities. The common marine protozoan *Noctiluca* emits light in response to shock and can be present in sufficient numbers to be easily visible. This phenomenon is taken up further in the next chapter. Finally, sonoluminescence may be involved as the sea water is alternately compressed and expanded (see below).

Fernand Montessus de Ballore studied South American earthquakes and uncovered reports of sparks from telegraph and power lines during an earthquake. The precise cause was not ascertained although a similar observation has been reported during a Mexican quake. The breaking of a high tension cable could be responsible as it would result in arcing for a short period of time. For a 250 kV cable breaking under the force of gravity the flashes would last for about 0.13 seconds at 50 hertz. Flashes could also occur upon grounding but in all cases they would not light up the sky. In addition several earthquake light reports date back well before the advent of power cables.

6.5.2   Earthlights

There are numerous reports of unidentified lights associated with prominent structures such as mountains and tall buildings throughout the world. Particular interest relates to places where the phenomena have been seen repeatedly. Here they have often sparked interest in the popular press and been the subject of several books. They have been described collectively as 'earthlights'. These phenomena are diverse and many of the accounts are unsubstantiated. Three well-researched examples are given. Others can be found in the further reading section below.

6.5.2.1   The Egryn Lights.   A small collection of houses and a chapel of this name are situated close to the North Wales coast about 12 km south of Harlech. In early December 1904, a large arc-like aurora was reported spanning the sky from the mountains in the east to the sea in the west. A few weeks later three people see a large light to the south of the chapel followed by several similar sightings by different people extending until the end of January. London journalists heard about the sightings and hurried to the area for a story. Throughout the following February and March more lights were seen both around Egryn and further afield.[21]

Descriptions of the lights varied. A Daily Mail reporter saw 'very bright' lights above Egryn chapel from a mile distant when a service was in progress. A yellow ball of fire was suddenly seen above the chapel roof, lasting about one and a half minutes before being extinguished. Blue, red and white lights were also reported, some as streaks, others as globes or glowing bars. Some were faint and aurora-like, others more intense and bright. Many suggestions were made at the time, the more popular being St Elmo's Fire and the *Ignis fatuus*. These were dismissed as being too faint or being too far removed from wet ground. Also there was no record of stormy weather at the time. Another suggestion was the Northern Lights. At least some of the reports probably relate to them. One anonymous reader claimed they were sometimes seen in North Wales. More significantly, the period of observation coincided with the maximum of Solar Cycle 14 and large sunspots were reported at the time. One observer mentioned a large meteor being seen during the period, and was dismayed to find that no photography had been undertaken. Interestingly the Daily Mail reporter described the phenomena he saw were like 'large and brilliant motor car lights'. It is not clear where the reporter was standing at the time but there is another description from an observer who was walking on the hillside from Dyffryn to Egryn who saw a ball of light above Egryn chapel from about a mile away. At this point, the Harlech to Barmouth road behind the chapel is approximately aligned with his position so a vehicle headlight may have been visible.

By the end of 1904 there were about 23 000 motor cars in Britain. Electric lights were introduced into cars during the same year but most lights would have been of the acetylene type. These lights consisted of a bright yellow incandescent acetylene flame housed in front of a mirror. The resulting beam was poorly focused and would have spread out over a large area so easily seen.

The location of these incidents is of interest since they lie close to the Mochras Fault. The area was first surveyed geologically in 1855 but the presence of this fault was not considered significant at the time. As the association between earthquakes and light became better known, attention was directed towards faults in general. The Mochras Fault has attracted some attention as it forms the eastern boundary of the Tremadoc Bay Basin

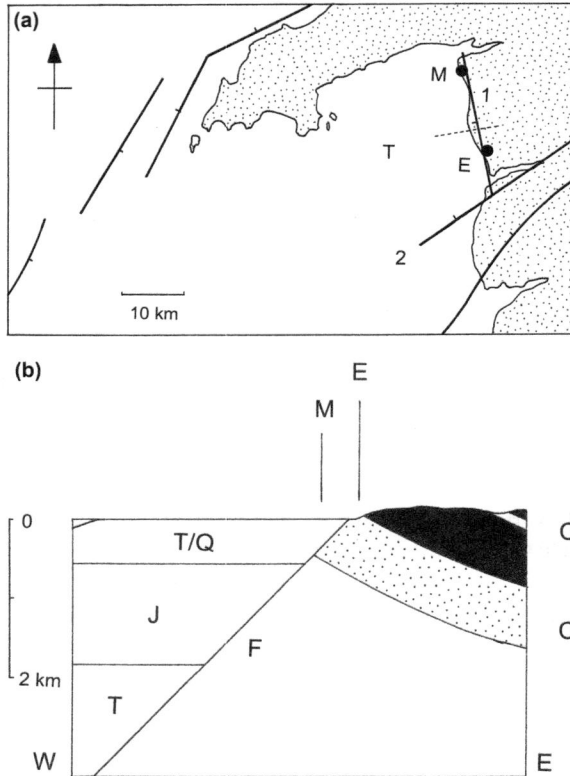

Figure 6.7 The Mochras Fault of North Wales. (a) General map showing the positions of faults. Land stippled. Broken line shows the position of the section in (b). 1 Mochras Fault; 2 Mawddach Estuary Fault; E Egryn Chapel; M Morfa Harlech; T Tremadoc Bay Basin. (b) Section through the Mochras Fault with equal vertical and horizontal scales, showing positions of the two 'earthlight' localities. C Sediments of Cambrian age; J Jurassic; T/Q Tertiary/ Quaternary; T Triassic. Plotted from maps of the British Geological Survey.

(Figure 6.7a and b). It has an exceptionally large throw of about 5 km resulting in the accumulation of large volumes of Mesozoic sediments on the seaward side. Although North Wales is a moderately seismically active region, no significant earthquakes have been attributed to this fault, although the presence of Quaternary sediments on the seaward side shows that the fault has been active in the geologically recent past. Whether its presence has any significance with respect to the Egryn lights remains questionable. There has been another unexplained occurrence close to this fault. It is the account of an unusual fire caused by a 'fiery exhalation' that burned several hayricks near Tyddin Sion Farm at Morfa Bychan.[22] It was communicated to the Royal Society by M. Lister in 1694. The circumstances are unusual and Lister ascribed the fire to an *ignis fatuus*. A more mundane cause might be the self-combustion of hay. Again many suggestions have been made including evasion of ionised gases and presence of strong electric fields but none with any basis in fact.

6.5.2.2 Hessdalen, Norway. Unidentified lights have been seen in this small valley, south of Trondheim since the 1930s but reached a climax in the early 1980s with numerous reports resulting in widespread media coverage. A serious attempt has been made here to record the lights and undertake a range of physical measurements to allow identification, beginning with Project Hessdalen in 1983. This was followed in 1998 by the construction of the Hessdalen Automatic Measurement Station. Since then, monitoring in one form or another has continued plus the publication of several detailed reports, with some in scientific journals.[23,24]

The lights last for a few seconds up to an hour or more. They come in several forms; spherical/circular, cigar-shaped, 'inverted Christmas tree' and triangular. They are frequently coloured, mainly red or yellow, stationary or with a jerky movement and sometimes flashing. They sometimes occur in multi-coloured groups.

The Hessdalen valley is about 15 km long running approximately north to south between two mountain ranges. It is traversed by the Hesja River and a road, with small communities scattered along its length. The valley floor is a mosaic of arable fields and pine forests rising up to bare mountain tops. The geology of the area is complex, consisting of a large sheet of basalt, with some gabbro, volcanics and an area of slate and phyllite (a coarser-grained slate). The Hessdalen research group has mapped what it considers as around 130 reliable sightings in the valley and surrounding mountains and found that the majority fall into an elliptical area spanning the valley towards the village of Ålen. The majority of the sightings occur in the northern sector of the valley between the mountains of Finnsåhøgda and Rogne rising about 400 m above the valley floor.

Despite much research into the geology, meteorology, magnetic anomalies and seismology of the area there is no convincing explanation for the lights although some have been confirmed as of automobile and aircraft origin. There is no shortage of explanations for the remainder but none of these appear particularly convincing. They include an 'earth battery' hypothesis where electricity is generated by deposits of copper on the Rogne with 'anode' rocks containing iron and zinc across the river while the acidic waters of the river itself complete the circuit. The magnetic field generated by the resulting flow of electricity causes the movement of the lights. There is also some correlation between the appearance of the lights and auroras. Earthquake lights have also been suggested. Mafic rocks such as basalt and gabbro when under stress become charged possibly leading to corona discharge and aurora-like phenomena, although this region of Norway is seismically quiet. Lightning strikes have also aroused interest and these have been mapped in detail. Most were of the usual downward negative type and were randomly distributed throughout the area. There was no significant correlation with sightings although the latter are most commonly recorded on foggy or cloudy nights. Surprisingly, spectra from some of the lights have revealed emission lines from silicon, iron and scandium. A connection with ball lightning was noted since this too may yield lines from silicon and iron. An interesting hypothesis has been presented by physicists.[25] They consider that the emission is occurring in a dusty plasma present in the atmosphere of this area. These plasmas contain small charged dust particles ($<10^{-6}$ m wide) and are thought to result from radon decay in the suspended dust. An electric field in the surface rocks below might then accelerate electrons within the plasma. The resulting fast electrons are decelerated by plasma ions and in doing so emit thermal bremsstrahlung (see Chapter 1). Some evidence for this is provided in the form of the Hessdalen lights spectra.[26] However, few measurements on the lights have been made with any precision. Nor is there any evidence of significant radon emission or local electric fields.

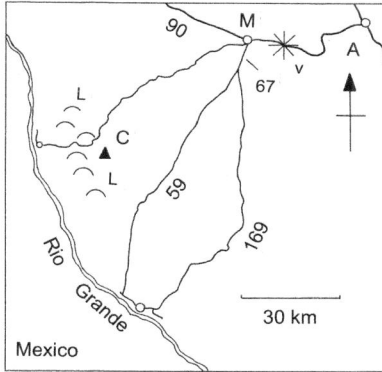

Figure 6.8 Map of the Marfa Lights district, Presidio Co., Texas. A Alpine; C Chinati Mountains, LL area of lights; M Marfa; v viewpoint. Route numbers shown.

6.5.2.3 Marfa Lights. First mistaken for Apache campfires in the 1880s, the Marfa Lights of western Texas have attracted thousands of visitors over the years. Balls of bright light have been reported up until the present day, either singly or in groups, stationary or moving and usually yellow-orange in colour. The best place to see them is along Highway 90 a few miles east of Marfa in Presidio County. They are most often reported in the direction of the Chinati Mountains about 80 km to the southwest (Figure 6.8). Suspicions arose early that they were distant car headlights as a number of roads cross the Chihuahuan Desert south of Marfa. The landscape is open with cactus and sagebrush scrub growing upon Permian sediments and volcanics. The land surrounding Marfa is at an altitude of about 1400 m, rising to over 2000 m in the Chinatis. Visibility is usually good in this region and temperature gradients over the flat landscape are likely to produce mirages. The bending of distant lights caused by temperature gradients in the atmosphere has often been suggested as a means by which lights may be seen in unexpected places.

There have been several serious studies of the lights.[27,28] Monitoring stations were organised to analyse their frequency and direction and in the early 2000s a group of physicists undertook some instrumental and statistical studies. They found that the frequency of sightings from the view point looking towards the mountains correlated significantly with the frequency of traffic on Route 67. When a car parked there was seen to flash its lights it looked like a typical Marfa light. Spectral analysis of some of the lights also showed that they were probably from vehicles or small fires. Although not all of the Marfa lights can be explained in this way, these studies demonstrate the importance of presenting a reasoned hypothesis and testing it experimentally using appropriate equipment and expertise.

6.6   THERMOLUMINESCENCE

While experimenting with minerals, Robert Boyle discovered that light was given off by a diamond heated well below its incandescence point. He also warmed a diamond by taking it to bed and noticed a faint glow in the dark. In 1838 du Fay discovered that emission from heated quartz could only be sustained if it was exposed again to light and in the early 19th century it was found that quinine sulphate emitted a strong blue light when warmed to the boiling point of water. The phenomenon of non-incandescent light emission upon the heating of crystalline materials is called thermoluminescence. The term has been criticised on the grounds that it implies that heat is the fundamental cause of the emission which it is not, but the term is too well established to be changed. Thermoluminescence results from the interactions of solar radiation or natural radioactivity with electrons within materials. Some electrons are mobilised to the conduction band and fall into traps (see Chapter 1). Raising the temperature increases the crystal lattice vibrations which can lead to the release of the trapped electrons. These excited electrons are then capable of emitting light. At the atomic level the process can be regarded as a form of phosphorescence and several phosphorescent minerals are also thermoluminescent such as fluorite and tourmaline. In a book on experimental chemistry by G. Carey,[29] thermoluminescence was demonstrated by powdering a small quantity of Derbyshire fluorspar then throwing it onto a plate heated to just above the boiling point of water. In a darkened room the minerals gave a bright display.

A substance must possess three fundamental characters in order to thermoluminesce; it must be either an insulator or a semi-conductor; it must have absorbed some energy from a source of radiation and it must re-radiate some of this energy when heated.[30] The re-radiation is subject to the Stokes effect so that the wavelength of the radiated light is of lower energy than the absorbed light (longer wavelength).

Thermoluminescence has several useful applications. Archaeologists use it to date pottery.[31] The firing temperature of pottery is above 500 °C, sufficient to release the majority of electrons from their traps. The thermoluminescence activity of the pottery is effectively re-set to zero, but trapping will continue to occur once it has cooled *via* the natural radioactivity of the material itself and that of its surroundings. Samples recovered from archaeological sites therefore contain a 'clock' that can be measured by re-heating the material and measuring the amount of radiation that is emitted. The amount of radiation is directly proportional to the amount of time that has elapsed since firing. Calibrations are required such as the measurements of radio-activity within the pot and its surroundings. This is possible because radioactivity is not affected by heat.

6.7 SONOLUMINESCENCE

Light emission from bubbles was first described in experiments with sonar, used to detect underwater craft. Several hypotheses have been advanced for this emission which is now known to be complex involving several processes. Among the earlier suggestions, the 'hot spot theory' appeared the most plausible.[32] During the rapid collapse of a bubble, adiabatic heating would increase the temperature sufficiently for the emission of black-body radiation (incandescence). Owing to the fact that the emitting bubbles were microscopic, often less than 10^{-6} m wide and the short collapse and emission times, experimental studies were challenging and initially led to several incompatible findings. Most experiments were conducted in water into which a transducer was immersed to produce ultrasonic waves. Hydrophones and photomultipliers were put in place to simultaneously record bubble collapse and light emission as the waves alternately compressed and expanded the bubbles (Figure 6.9). Most studies concluded that the light emission occurred close to the collapse time giving credence to the hot spot theory. When water was saturated with different noble gases it was found that

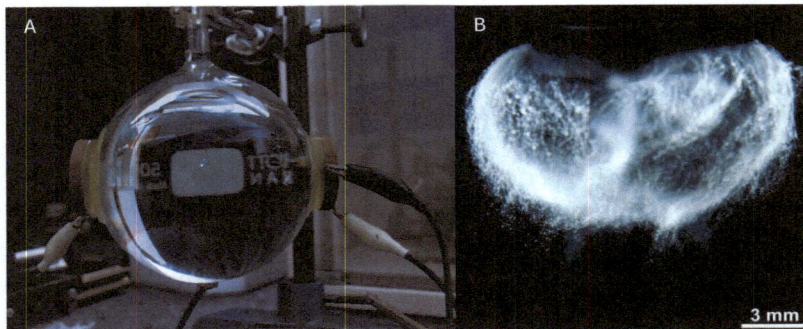

Figure 6.9 (A) Single bubble sonoluminescence (Reproduced from ref. 34, https://commons.wikimedia.org/wiki/File:Sonoluminescenza. jpg, under the terms of the CC BY 3.0 license, https:// creativecommons.org/licenses/by-sa/3.0/). (B) Multi-bubble sono-luminescence in 85% phosphoric acid saturated with xenon. Re-produced from ref. 35 with permission from John Wiley & Sons, Copyright © 2010 Wiley-VCH Verlag GmbH & Co. KGaA, Weinheim.

the more thermally conducting gases such as argon reduced the luminescence which also agreed with the theory. In addition, spectroscopic studies found that the light emission fitted the black body curves fairly well. But later more detailed studies using a range of experimental conditions and different gases revealed emission bands and lines that were clearly the result of other processes. At the high temperatures attained, gas molecules readily fragment into ions, including free radicals. Using air-saturated water, evidence of free radicals from nitrogen and water molecules in their excited state have been identified along with the excited molecules. Upon de-excitation and recombination, light is emitted. Studies have also identified brems-strahlung-like behaviour as fast particles enter the water immediately outside the bubble.[33]

Light emission has therefore been shown to be related to the nature of the bubble gases. Some researchers believed that they had discovered nuclear fusion in 2006 during cavitation experiments. Their work has not been confirmed and fusion is considered to be most unlikely.

Cavitation and sonoluminescence are not restricted to the laboratory. Pistol shrimps produce a faint sonoluminescence by snapping their claws. A small bubble is formed and the process

Figure 6.10   Line drawing of a pistol shrimp showing the large cheliped at left. This is snapped shut using powerful muscles. Light is emitted during the event. Shrimp is 20 mm long.

leads to a shock wave that can stun nearby prey. Some mantis shrimps can also perform this action. The claws of these colourful shrimps are modified to produce this effect. They are common in the seas and are responsible for a significant amount of ocean 'noise'. An example of a pistol shrimp is shown in Figure 6.10.

FURTHER READING

A. Feng and P. F. Smet, A review of mechanoluminescence in inorganic solids: components, mechanisms, models and applications. *Materials*, 2018, **11**, 484. 10.3390/ma11040484.

S. Gleason, *Ultraviolet Guide to Minerals*, Van Nostrand Reinhold, USA, 1960.

H. W. Leverenz, *An Introduction to Luminescence in Solids*, Dover, New York, 1968.

S. Shionoya and W. M. Yen, *Phosphor Handbook*, Chemical Rubber Company, Boca Raton, 1999.

F. Turnbull, Fascinating fluorescence. *Gems*, 1973, **5**, 15–17.

A. J. Walton and G. T. Reynolds, Triboluminescence. *Adv. Phys.*, 1984, **33**, 595–660.

E. Yaffa and E. Shalom, *The Fourth State of Matter, An Introduction of the Physics of Plasma*, Adam Hilger, Bristol, 1989.
W. M. Yen and M. J. Weber, *Inorganic Phosphors: Compositions, Preparation and Optical Properties*, CRC Press, Boca Raton, 2004.

REFERENCES

1. S. H. Ball, Luminous gems, *Sci. Mon.*, 1938, **47**, 495–505.
2. G. Rose, *Mineralogisch Geognostische Reise nach den Ural, dem Altai und dem Kaspische Meere*, Sanderschen Buchhandlung, Berlin, 1837–1842, vol. 2.
3. H. E. Millson and H. E. Millson Jr., Duration of phosphorescence II, *J. Opt. Soc. Am.*, 1964, **54**, 638–640.
4. E. T. Holland, Notes and queries (phosphorescent snow), *Alp. J.*, 1863, 143–144.
5. M. Minnaert, *Light and Color in the Open Air*, Dover Publications, New York, 1954.
6. F. Licetus, *Litheosphorus Sive De Lapide Bononiense*, Bologna, Italy, 1640.
7. A. Roda, A History of Bioluminescence and Chemiluminescence from Ancient Times to the Present, in *Chemiluminescence and Bioluminescence: Past, Present and Future*, ed. A. Roda, Royal Society of Chemistry, Cambridge, 2011, pp. 3–50.
8. C. A. Balduin, Phosphorus hermeticus sive Magnes luminaris, *Misc. Acad. Nat. Cur. Dec. I.*, 1673–1674, 105–172 (5th year).
9. W. Homberg, Nouveau phosphore, *Mém. de l'Acad. Roy. des Sci.*, 1693, **10**(445–448), 1730.
10. J. Canton, An early method of making a phosphorus, that will imbibe and emit light, *Phil. Trans. R. Soc.*, 1768, **15**, 337–344.
11. D. Zitoun, L. Bernaud, A. Manteghetti and J.-S. Filhol, A microwave synthesis of a Long-Lasting phosphor, *J. Chem. Ed.*, 2009, **86**, 72–75.
12. J. Goettlicher, A. Kutelnikov, N. Suk, A. Kovalski, T. Vitova and R. Steininger, X-ray absorption near edge structure spectroscopy on the photochrome sodalite variety hackmanite, *Z. Kristallogr. – Cryst. Mater.*, 2013, **228**, 157–171.
13. O. Wässtrom, Über einer besondern Schein im Wasser der Ost-see oder das in der Scheren von Wermdoe Schwachfeuer, *Ann. Phys.*, 1799, **2**, 352–358.

14. R. M. W. Musson, *A catalogue of British Earthquakes*, British Geological Survey Technical Report WL/94/04, Edinburgh, 1994.
15. G. Neilson, R. M. W. Musson and P. W. Burton, *Macroseismic reports on historical British earthquakes X1. 1931 Jun 7 North Sea*, BGS Global Seismology Report No. 280, 1986.
16. C. Davison, *The Hereford earthquake of December 17, 1896*, Cornish Brothers, Birmingham, 1899.
17. S. Takaki and M. Ikeya, A dark discharge model of earthquake lightning, *Jpn. J. Appl. Phys.*, 1998, **37**, 5016–5020.
18. R. Thériault, F. St-Laurent, F. T. Freund and J. S. Derr, Prevalence of Earthquake Lights Associated with Rift Environments, *Seismol. Res. Lett.*, 2014, **85**, 159–178.
19. F. St-Laurent, J. Derr and F. Freund, Earthquake lights and the stress-activation of positive hole charge carriers in rocks, *Phys. Chem. Earth*, 2006, **31**, 305–312.
20. E. Roberts, *Our Quaking Earth*, Little, Brown & Co., Boston, 1963.
21. P. Devereux, *Earth Lights Revelation*, Blandford Press, London, 1989.
22. M. Lister, An Account of the Burning of several Hayricks by a fiery Exhalation or Damp, *Phil. Trans. R. Soc. Lond.*, 1694, **18**, 49–50.
23. G. S. Paiva, Hessdalen lights produced by electrically active inversion layer, *Meteor. Atmos. Phys.*, 2021, **133**, 1447–1454.
24. M. A. Teodorani, A long-term scientific study of the Hessdalen Phenomenon, *J. Scientific Explor.*, 2004, **18**, 217–251.
25. G. S. Paiva and C. A. Taft, A hypothetical dusty plasma mechanism of Hessdalen lights, *J. Atmos. Solar*, 2010, **72**, 1200–1203.
26. G. S. Paiva and C. A. Taft, A mechanism to explain the spectrum of the Hessdalen Lights phenomenon, *Meteorol. Atmoph. Phys.*, 2012, **117**, 1–4.
27. E. Darack, Unlocking the Atmospheric Secrets of the Marfa Mystery Lights, *Weatherwise*, 2008, **61**, 36–43.
28. K. D. Stephan, J. Bunnell, J. Klier and L. Kanala-Noor, Quantitative intensity and location measurements of an intense long-duration luminous object near Marfa, Texas, *J. Atmos. Solar*, 2011, **73**, DOI: 10.1016/j.jastp.2011.06.002.

29. G. G. Carey, *Five hundred useful and amusing experiments*, London, 1825.
30. S. W. S. McKeever, *Thermoluminescence in solids*, Cambridge University Press, 1985.
31. M. J. Aitken, *Thermoluminescence Dating*, Academic Press, London, 1998.
32. J. Rooze, E. V. Rebrov, J. C. Schouten and J. T. F. Keurenjes, Dissolved gas and ultrasonic cavitation. A review, *Ultrason. Sonochem.*, 2013, **20**, 1–11.
33. W. C. Moss, D. B. Clarke and D. A. Young, Calculated pulse widths and spectra of a single sonoluminescing bubble, *Science*, 1997, **276**(5317), 1398–1401.
34. F. Lembo, File: Sonoluminescenza, available from: https:// commons.wikimedia.org/wiki/File:Sonoluminescenza.jpg.
35. K. S. Suslick, N. G. Glumac and H. Xu, *Angew. Chem., Int. Ed.*, 2010, **49**, 1079–1082.

CHAPTER 7

Luminous Seas

7.1 INTRODUCTION

We all followed his movements with our eyes, for undoubtedly some nervousness was growing on us, and we saw a whole mass of phosphorescence, which twinkled like stars.

Dracula chapter XIX by Bram Stoker

The impracticality of considering the whole of bioluminescence has been narrowed down to one topic, namely 'phosphorescent' seas which deals with microscopic bioluminescing planktonic organisms. This will be compared with bioluminescence in land plants and terrestrial microbes in Chapter 8.

Bioluminescence is caused by the emission of light from a biomolecule called luciferin. Luciferins are large molecules containing aromatic ring structures and are found in a wide variety of animals, some fungi and bacteria. Several different molecules are known to fulfil the role of luciferin and while some organisms contain the same molecule, in others the molecule appears to be unique to the species. Luciferin has to be oxidized with molecular oxygen before it can emit light and the necessary chemical reaction is catalyzed by the enzyme luciferase. An electronically excited intermediate causes the emission as its

Luminous Phenomena: A Story of Spontaneous Combustion, Phosphorescence and Other Cold Lights
By Allan Pentecost
© Allan Pentecost 2025
Published by the Royal Society of Chemistry, www.rsc.org

electrons fall back to the ground state. Complications arise because luciferases also differ between different organisms so the term represents widely differing substances that are capable of essentially the same process. Today not all of the structures of these compounds are known, nor their precise mode of action.

Light emission by organisms will have been recognized by the earliest humans but it first appears in the Western literature in the 4th century BCE when Greek writers mentioned the glowing of rotting wood and dead fish. Little progress was made until Robert Boyle found that 'glow worms', the larvae of small European beetles ceased glowing if air was excluded from their surroundings. A few years later the Italian scientist Marcello Malphigi turned his recently-developed microscope onto fireflies and discovered that small granules within the cells were responsible for the emission. During the Age of Enlightenment which stimulated much scientific experiment and thought, little further progress ensued although Spallanzani experimenting with luminous jellyfish and shellfish found dried preparations of these animals stopped luminescing but continued to do so once the material was moistened. Finding that the luminescence is usually lost when the organism was placed in gases such as nitrogen and carbon dioxide led to the (incorrect) conclusion that the element phosphorus was involved since it too loses its glow in these gases. Scientists were unable to explore the subject further until the true nature of light and chemical combination was understood and this did not occur until later in the nineteenth century.

In 1885 a breakthrough was made by Raphael Dubois who prepared a paste out of a luminous clam and observed the effects of water and temperature.[1] By this means he was able to show that a chemical reaction was involved and consisted of at least two components. One component was sensitive to heat but the second component was not. Dubois succeeded in a partial isolation of the two substances and named them luciferase and luciferin, respectively. Enzymes such as luciferase are a group of substances known for their intolerance of heat.

Research on incandescence had already shown that the light emission obeyed a physical relationship that provided a relationship between light and temperature but was soon seen to be inappropriate for bioluminescing organisms. As a result, Eilhard Wiedermann defined this new 'cold light' as *chemiluminescence* in 1888. The term *bioluminescence* was first used by Mangold in 1910. It is essentially a form of chemiluminescence. Once these discoveries became known to the wider scientific community research continued apace. Early in the 20th century the concept of 'active oxygen' appeared and advances in light-measuring instruments allowed very weak light emissions to be detected. Ultimately it became possible to purify some of the luciferins and luciferases allowing their structures to be determined. This led to more detailed experiments when it became clear that the molecular structures of both components varied considerably from organism to organism. In recent years luciferase has played an increasingly important role in molecular biology and genomics. Luciferase genes – those sections of an organism's DNA responsible for the synthesis of the luciferase protein can be attached to another gene – a 'gene of interest', using the technique known as *gene fusion*. The resulting entity can be inserted into small circular molecules of DNA known as plasmids. Plasmids can be inserted easily into cells where they can function as part of the cell's total DNA complement. The 'genes of interest' will be those that are difficult to detect within cells using conventional methods but by adding luciferase their presence is betrayed by their bioluminescence. Only cells containing the altered DNA will respond in this way. It is not the DNA itself that is being targeted but the protein expressed from the original gene fusion. The technique only works if the organism's DNA contains no indigenous luciferase genes but has proved valuable in detecting the activities of particular genes within cells. The gene responsible for producing the luciferase protein is known as a *reporter gene.*

7.2   EARLY HISTORY

'In three days the sun shall shine'
'On our return – till then all peace be thine'
This said, his brother Pirate's hand he wrung
Then to his boat with haughty gesture sprung

Flash'd the dipt oars and sparkling with the stroke
Around the waves' phosphoric brightness broke

Byron: *The Corsair* Canto 1 lines 565–70 (1814)

Bioluminescence in the sea was recorded by the Greek writer Anaximenes who noted that emission of light occurs when oars struck the water. According to E. Newton Harvey in his detailed history of the topic[2] this is the earliest account currently known (*ca.* 500 BCE). Fishermen would have been well acquainted with the light, since fish shoals can sometimes be located by them. Despite the widespread occurrence of marine bioluminescence, surprisingly few accounts have come down to us from the classical period and regular descriptions only begin to appear in the 16th and 17th centuries. By 1666 it became sufficiently well reported in scientific circles that the Royal Society offered the '*sea shining at night*' as an appropriate topic for research and in the following year Sir Robert Moray published a brief description of the phenomenon. Whilst staying at Deal in Kent, he reported that sailors were well acquainted with the 'phosphorescence' as it was then known, and associated it with easterly and south-easterly onshore winds. Moray was travelling to Jamaica and there he found that the light was influenced by water currents as well. There was however, little to add to that already known. It was understood that sea water, when disturbed can emit tiny sparks of light for a fraction of a second, but at a time before the development of good microscopes, their origin could not be ascertained since most of the organisms responsible were not clearly visible.

Despite the lack of an experimental approach, there was no shortage of hypotheses accounting for the light. Among them was the suggestion that electricity was responsible and the eminent scientist René Descartes thought that the salt of the sea, leaving as fine spray became fragmented and emitted light. After all, freshwaters were not luminescent, and some sea creatures such as the electric eel produced an electric current. Other observers thought the light came from the sun either through its heating effect or through a delayed re-emission. It was argued that because on calm and bright days, when water was collected at a depth of about 15 m it showed little sign of luminescence, but luminescence increased if the water was placed in the sun for a few hours. Domenico Bottoni regarded bioluminescence as a form of fire induced by motion, referring to the bright sparks emitted by marine organisms.[3] Likewise, Isaac Newton thought that some kind of agitation was responsible citing the glow of seawater in a raging storm.

Small advances were nevertheless made in favour of a biological origin. In the early 1700s Father Bourzes a Jesuit missionary noted that the sea glow was particularly bright in areas where the water was discoloured red or yellow and had a glutinous consistency. When luminous sea water was strained through a cloth, James Bowdoin found that the luminosity was reduced showing that small suspended particles of a possible biological origin were responsible. Benjamin Franklin with whom Bowdoin was associated, agreed with this suggestion.

7.3 ORGANISMS RESPONSIBLE – DINOFLAGELLATES AND BACTERIA

It is now known that the organisms responsible are part of the plankton, a collective term for those aquatic inhabitants unable to propel themselves against a current, thus distinguishing them from fishes and marine mammals. They include bacteria, algae, protozoa, crustaceans, jellyfish and fish larvae. Members of all these groups have bioluminescent species and the first to be discovered was a dinoflagellate protozoan.

Dinoflagellates are a group of microscopic organisms dispersed throughout the oceans. They are unusual plankters and have been claimed by both botanists and zoologists. Many of the species possess chloroplasts and are able to photosynthesise, making them candidates for the plant kingdom as algae, while others obtain their energy solely through the capture and ingestion of other organisms making them protozoa, a group of simple animals. Even some of the 'algal' members are known to capture and ingest prey. They are best regarded as protozoa, some of which have probably 'stolen' chloroplasts from other algae to provide an alternative or additional source of energy, as a form of symbiosis. Organisms such as these form diverse assemblages in the oceans and have recently been described as mixoplankters, combining their ability to photosynthesise and ingest organisms and other organic particles as food. Taken as a whole, the dinoflagellates represent a major group of organisms responsible for much of the ocean's primary production upon which virtually all other non-photosynthetic marine organisms depend. They consist of small single cells almost all of which are equipped with two swimming organs known as flagella whose insertion into the cells gives them a unique shape. While many species have a thin and smooth cell wall Figure 7.1a), others are covered with stiff angular plates (Figure 7.1b). Dinoflagellates have other unusual characters. Their nuclei are exceptionally large for their size and the structure of their chromosomes is unique among organisms. Equally remarkable is the fact that while dinoflagellates are often abundant in fresh water, none of these types are known to be bioluminescent.

Dinoflagellates are most numerous and often possess larger cells in coastal waters. Smaller dinoflagellates are more often encountered in tropical waters and in the open sea. In cooler seas, their abundance is seasonal and they sometimes betray their presence as dense coloured blooms during the warmer months of the year, an observation made long before their true nature was understood. In all about 80 different species are known to be luminescent, a small proportion of the whole. Prominent among them are the armoured luminescent forms

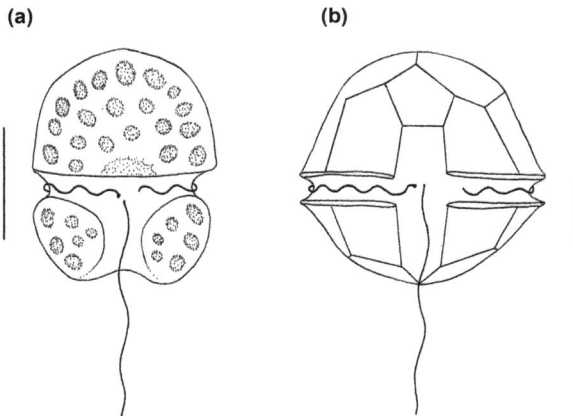

(a)   **(b)**

Figure 7.1   Dinoflagellates. (a) *Gymnodinium,* a naked species and (b) *Peridinium,* an armoured species. Note the two types of flagella. Chloroplasts are shown stippled in (a). Both genera have bioluminescent species. Bar 10 μm.

Tripos muelleri (formerly known as *Ceratium tripos*) and *Lingu-lodinium* (*Gonyaulax*) *polyedra*, but the large unarmoured *Nocti-luca* is by far the most significant, at least in the coastal waters and estuaries of warm and temperate seas.

Noctiluca was the first dinoflagellate to be discovered. In his fascinating book *Employment for the Microscope* Henry Baker in a chapter on luminous water insects[4] wrote that he had received a letter from Joseph Sparhall who hailed from the seaside town of Wells in Norfolk (Figure 7.2). Joseph wrote that '*animalcules which cause the sparkling light of seawater may be seen by holding a phial up to the light resembling very small bladders or air bubbles*'. Although his accompanying drawing has been lost, there is no doubt that Sparhall had seen the dinoflagellate that is now known as *Noctiluca* (Figures 7.3 and 7.4). What is most remark-able is the fact that this large species can sometimes be seen with the naked eye since the cells range from 0.1–2 mm in diameter. They nevertheless escaped the attention of many a keen observer over the previous centuries including the famous microscopist Antonie van Leeuwenhoek. He was the first person

Figure 7.2   Wells by the Sea, home of Joseph Sparhall.

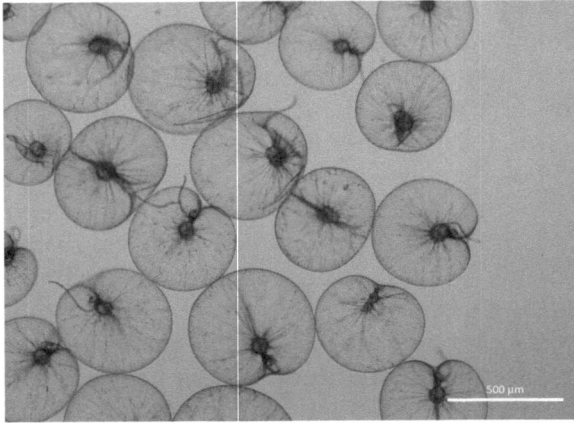

Figure 7.3 *Noctiluca* cells showing central nucleus (dark) radiating cytoplasm and tentacles. Reproduced with permission from Claire Widdicombe, Plymouth Marine Laboratories.

(a) **(b)**

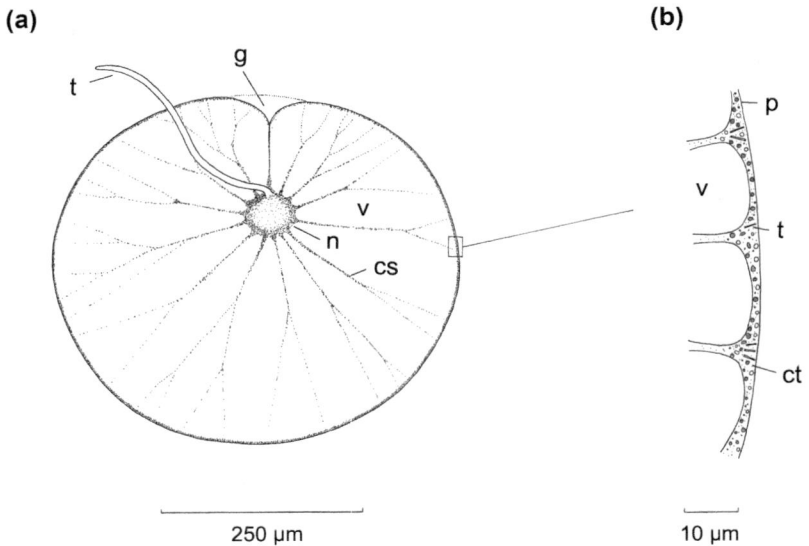

250 µm 10 µm

Figure 7.4 (a) Diagram of an entire *Noctiluca* cell showing branched cytoplasmic strands connecting the nucleus to the peripherial zone (cs), tentacle (t), cell groove (g), nucleus (n) and flotation vacuole (v). Bar 250 µm. (b) Detail of the cell peripheral zone showing thin cytoplasm (ct), outer membrane or plasmalemma (p), trichocysts (t) and flotation vacuole (v). Dark bodies within the cytoplasm are the microsources. Bar 10 µm. Trichocysts are organelles capable of rapid projection and serve a protective function.

to describe the protozoa and lived less than 10 miles from the sea. Experiments with *Noctiluca* soon followed. It was found that if water containing it was shaken continuously the emission of light gradually diminished but if the water was left to rest for a while the luminescence was eventually restored. Further microscopy revealed that most light emission occurred close to the surface of the cells. M. Rigaut in 1765 made a careful study of the effects of chemicals on *Noctiluca* emission. In some cases adding dilute acids to the sea water increased the luminescence and the same effect could be achieved by adding ammonia but no explanation could be found. James Macartney named the organism *Medusa scintillans*[5] and in 1816 it first appeared in print as *Noctiluca miliaris* later being corrected to *N. scintillans*. Macartney had suggested that the emission of light could have a protective function for this organism. Early microscopy was bedeviled by the lack of clear images, especially at the higher magnifications and it was not until the 1820s when achromatic lenses were invented that things began to improve leading to many new discoveries.

Work was sufficiently well advanced by 1850 that the French biologist Jean Quatrefages wrote a monograph on *Noctiluca* and suggested that the light originated from bodily contractions.[6] He produced good illustrations of the organism and showed that it emitted a steady glow of light under some conditions and confirmed Macartney's observations that the flashes of light originated from small particles in the protoplasm. Other work had shown that probing *Noctiluca* with a needle stimulated light emission as could a change in the salt content of the water. During this period when exploratory expeditions were taking place to lesser known parts of the world, many descriptions appeared of bioluminescent seas. Charles Darwin described in his journal the bright luminosity of wave crests which could be so extensive as to be reflected from the sky. There are also more recent accounts of luminosity appearing along the crests of tsunamis most likely resulting from dinoflagellates such as *Noctiluca*. Bioluminescence of sea water was found to be a common phenomenon occurring in all but the coldest seas and resulted from the agitation of the water.

Interest in dinoflagellate luminescence continued into the early 20th century although progress was held back to some extent by the two world wars and a lack of advanced technology particularly in the field of microscopy and biochemistry. Ironically some new discoveries were made as a direct result of the wars. During WW1 an observer at Lowestoft on the East Anglian coast saw a zeppelin dropping bombs a few miles out at sea. On exploding, the shock waves caused a flash of light along the shore. During WW2 a submarine was reported to have been destroyed as a result of its tell-tale passage through bioluminescent sea water and during the Pacific conflict much research was conducted into the visibility of allied warships to enemy planes as they passed across areas of luminescent plankton. Ship's radars have occasionally been reported to influence the luminescence close to them but it remains unclear how this could occur. Ship's radars emit microwaves in the 7.5–15 cm region. They are strongly absorbed by water at around 12 cm (principle of the microwave oven) and dependent upon the roughness of the sea, considerable radar absorption could take place. The resulting increase in water temperature would be extremely small and although this would increase turbulence any effect on the surface plankton is likely to be minimal.

In more peaceful times some attention was diverted to the problem of *Noctiluca* flotation. Joseph Sparhall, mentioned above, had first observed that the cells contained within his phial tended to rise to the surface of the water when left undisturbed, implying that they were less dense than sea water. But how was this achieved? The cells contain some oil that is lighter than sea water but not enough to keep the cells afloat. *Noctiluca* is not a typical dinoflagellate as it possesses a huge vacuole that accounts for most of the cell volume. It was later found that the cells were isotonic with sea water, that is to say, they contained the same molecular concentration of solutes as sea water in their vacuolar cell sap (Figure 7.4a). But some molecules, and especially ions that are commonly found within vacuoles, are heavier than others so it was suggested that the vacuole selected the lighter ions to help them float. Experiments with other dinoflagellates have since shown that by excluding the heavier ions of calcium, magnesium and sulphate from the vacuole's cell sap, and replacing them with the lighter ions of sodium brings dinoflagellate cells close to neutral buoyancy.[7] The lightest ion is that of hydrogen but this will affect the pH (acidity) of the water. However further measurements on the *Noctiluca* vacuole indicated a pH of around 3.5 demonstrating a significant concentration of hydrogen ions was present as well. This no doubt makes a further contribution to *Noctiluca's* ability to float.

During the early 20th century advances were also made in understanding the morphology and cytology of *Noctiluca* cells. This organism is unusual among dinoflagellates in possessing a peculiar flagellum often described as a tentacle in addition to the second flagellum. The tentacle is not an efficient means of propulsion but has been adapted for capturing prey. The prey is transferred to a layer of sticky mucilage before being ingested.

It had been known from some time that the light source in *Noctiluca* was confined to the thin layer of cytoplasm located next to the cell's outer membrane. Powerful microscopes revealed that a large number of small particles were responsible for the emission and in a related dinoflagellate called *Lingulodinium* they were termed *scintillons*[8] although the more general term *microsource* has subsequently been used since there is some structural variation between different dinoflagellates.[9] The microsources of *Noctiluca* are minute membrane-bound structures occurring within the cytoplasm but associated with the vacuole and contain luciferin and luciferase (Figure 7.4b). By the 1960s it became possible to grow luminescent dinoflagellates in the laboratory and extract many of the active components by breaking open the cells and subjecting the contents to centrifugation. By this means crude extracts of the microsources could be examined more carefully and some of their properties determined. However it was still not clear how the disturbance of sea water resulted in the emission of light. Experiments had already shown that altering the chemistry of the water or by passing a small electric current through the cells resulted in emission but this is far removed from the physical disturbance of the water. The answer was initially found through studies that had been undertaken in nerve cells whose activity depended upon an action potential (see Chapter 1). Special proteins embedded in the cell's membrane respond to small changes in pressure by altering their structure. This leads to a cascade of biochemical events opening up ion channels within the membrane and changing their electrical properties. The resulting electrical signal is transmitted rapidly and is picked up by the microsources. Here chemical and structural changes within the luciferin–luciferase

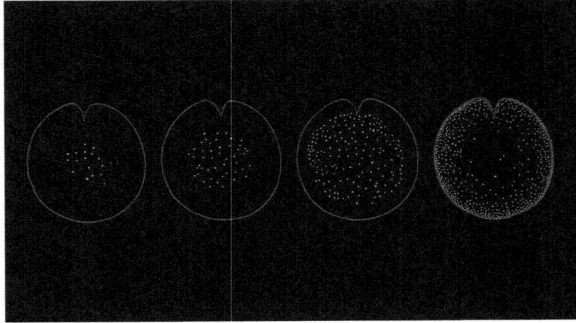

Figure 7.5 Diagram showing the sequence of light emission by micro-sources in *Noctiluca*. From left to right, a cell is stimulated into emission and the action potential spreads around the cell's periphery stimulating further sources as it goes. A point of maximum emission is rapidly achieved after which there is a slower decline. Note that only a small number of the micro-sources are shown in the diagrams.

system cause the oxidation to proceed with the emission of light.[10] Each flash has been found to release up to 800 photons of visible radiation. The whole process occurs within a fraction of a second and the signal is rapidly transmitted throughout the cell containing thousands of microsources (Figure 7.5).

Noctiluca emission suffers from what is sometimes referred to as 'fatigue'.[11] Continued stimulation soon exhausts the luminescence but a partial recovery occurs after a few minute's rest. Oxygen is an important component of this system and this is easy to overlook, but close to the sea surface it is in plentiful supply.

7.4 *NOCTILUCA* ECOLOGY AND DISTRIBUTION

Noctiluca is common in the North Atlantic and has been reported on numerous occasions. Its luminescence at night makes for easy detection even in low numbers and there are several studies of its absolute abundance in blooms. In the Black Sea numbers have been found to range from 20–7200 per cubic metre but in other regions they can reach higher values – up to 100 000 per cubic metre have been reported in the English Channel with an excess of three million cells per *litre* in a sample collected from the Gulf of California. A recent study using satellite imaging revealed a vast bloom of *Noctiluca* in the Beibu Gulf of southwest

China.[12] The area covered more than 20 000 km^2 and the bloom appears to have been influenced by the upwelling of plant nutrients along the Chinese shore. The nutrients would have stimulated phytoplankton growth, providing a food source for the *Noctiluca*. This bloom lasted about 15 days and was abruptly dissipated. Wind disturbance was clearly a factor – when the wind speed fell, the surface bloom became more apparent, most likely the result of flotation when the sea water was less turbulent. This dinoflagellate feeds largely upon diatoms, a group of planktonic algae that undergo their own 'blooming' in the spring or early summer. The reddish colour of many *Noctiluca* blooms said to resemble watered down tomato sauce is due to carotenoid pigments within the cells but there is also a green form of *Noctiluca* that is confined to the warmer seas of the tropics. Blooms of *Noctiluca* often follow the diatom bloom but may persist through most of the summer when diatoms are scarce indicating that other plankters are also used as food. This is amply borne out by microscopic and culturing studies. For example, it is known that this organism will feed on other dinoflagellates and small animals. In North Wales it was found that oyster embryos were avidly consumed leading to concerns over the local oyster beds. Cultured *Noctiluca* can also be fed on a range of microalgae in addition to diatoms.

Noctiluca is normally only found at or near the sea surface. It is uncommon below about 25 m depth where planktonic algae occur in smaller numbers owing to the diminution of light which they need to survive. There is no evidence that it possesses a daily rhythm of light emission although this has been found in some other dinoflagellates.

It has been seen that there is plenty of lore associated with bioluminescent seas and these must have been a common sight in the days of sailing boats whose slow passage allowed plenty of scope for sailors to examine the water they were sailing through. Nevertheless interesting and sometimes amusing reports continue to emerge. A colleague told me of an incident along the Humberside coast. At the time he was on pollution duty for the Environment Agency. One summer's night he was rudely awakened by a telephone call at 3 am. A group of inebriated lads had decided to go for a midnight splash fully clothed in the sea. They were worried because when they emerged there was a peculiar glow in the water, and a nuclear power station was operating nearby. After having already had a long and busy day and with three young children at home my friend was not at all pleased. He drove down to the coast, gave them a worried look, mentioned the power station and told them that samples would need to be taken. Tests would have to be made and the results would not be known until the following day. Of course, he knew all along that the culprit was *Noctiluca*!

So why are some dinoflagellates luminescent? Strongly in favour is the suggestion that it will deter potential predators. A group of small crustaceans, the calanoid and cyclopoid copepods are likely the most significant predators of dinoflagellates as they feed upon planktonic algae and protozoa and are common and widespread in the oceans. The planktonic calanoids are filter feeders and create a current of water using their appendages to guide food particles toward their mouths. Cyclopoid copepods tend to be direct feeders, catching prey using their mouthparts. Both methods of feeding generate water disturbance and of course direct contact between predator and dinoflagellate will elicit a luminous response. Copepods would be able to detect the emission since they can see although their single eye can only detect the general direction of light emission and focusing is not possible. They feed close to the sea surface at night to avoid their own predators who have better vision. Feeding on luminescent dinoflagellates the copepods themselves with their translucent bodies could also light-up and this, combined with the general disturbance caused by their feeding would provide a visual cue for predatory fish, squid and carnivorous copepods. Some experiments have supported this suggestion but it cannot be the whole story. For example, some planktonic copepods are regularly self-luminous and among the dinoflagellates the majority of species, including all of those living in fresh water, are consistently non-luminous. Even some populations of *Noctiluca* have been found to be non-luminous but the reasons for this are unknown. Luminous dinoflagellates are also known to possess a 'background' luminescence. This occurs when emission occurs without any apparent stimulation. There have been few studies of the phenomenon but it is likely to result from several causes: small organisms may come into contact with the cells without being observed or local turbulence and random molecular motions may also be responsible.

Conditions exist where marine species can penetrate several miles inland. Along low-lying coasts such as that of East Anglia, salt water is known to penetrate far inland under a layer of fresh water, owing to its greater density. Blooms of *Noctiluca* and other luminous dinoflagellates could light up creeks and ditches. In fact there are two 19th century reports of such occurrences. One was in the River Lothing near Lowestoft in June 1831 and another in the Rive Bure at Yarmouth sometime before 1884 although no microscopic examinations were made. The same area was once well known for its stories of Will o' the wisp' (Chapter 9).

7.5   OTHER LUMINOUS DINOFLAGELLATES

The geologist John McCulloch described what was probably a luminous *Cercaria (Ceratium)* from the Western Isles of Scotland in 1821.[13] In a voyage from the Mull of Kintyre all the way to Shetland he found luminous microscopic organisms almost every time he dropped the ship's bucket into the sea. Numbers often exceeded 100 to the cubic inch (6000 cells per litre). He found them to be particularly prevalent in the small harbours of Orkney and Shetland where he conducted some detailed observations of these and other luminous organisms in the summer of 1819.

In his great work of 1834 the German protozoologist Christian G. Ehrenberg (1795–1876) described and illustrated several further luminous dinoflagellates in sea water samples sent to him from Kiel. Currently the approximately 80 species of luminescent dinoflagellate are distributed among 18 genera. Apart from *Noctiluca*, most species belong to the genera *Alexandrium*, *Lingulodinium* (*Gonyaulax*), *Protoperidinium*, *Pyrocystis* and *Tripos* (Figure 7.6). Luminous dinoflagellates occur throughout the oceans but are best seen in warmer coastal seas. Several bays of the Caribbean are famous for their magnificent displays, particularly those of Puerto Rico but sightings around the British coast are also fairly common during warm summers and can be spectacular. They have created a considerable number of photography enthusiasts linked by social media (Figure 7.7). In most cases the organisms responsible are not known although *Noctiluca* is likely to feature strongly.

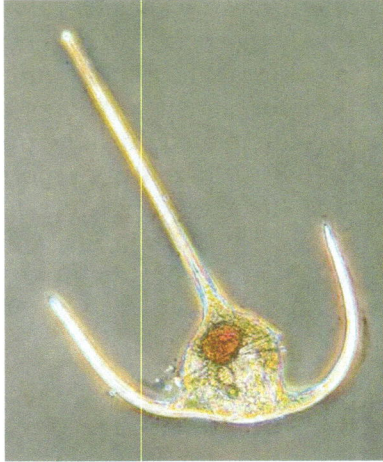

Figure 7.6   Image of *Tripos* (*Ceratium*) *muelleri*. The genus *Tripos* is characterised by the strong 'horns' on the cells. The species grow in both
fresh- and saltwater and are cosmopolitan. Cell is about 200 μm
long. Reproduced with permission from Mike Guiry and Robin
Raines (University of Galway).

Some detailed studies undertaken by Paul Tett on the Isle of
Cumbrae in the Firth of Clyde have identified the presence of
non-photosynthetic but luminescent *Lingulodinium, Peridinium*,
and *Polykrikos* species close to the shore.[14] These dinoflagellates
were investigated by placing samples of sea water in a thermos
flask and examining the water with a photomultiplier. Placing
them in the flask prevented exterior light contamination and
kept the water temperature reasonably constant and allowed the
'background' luminescence to be measured. When a dinoflagellate flashed, the signal would be detected by the photomultiplier and amplified to give a spike on a chart recorder. The
dinoflagellates in the sample were subsequently filtered off and
examined microscopically enabling their numbers to be estimated. In this way it was possible to assess their rate of flashing
in the flask. Measurements were made over periods of 15 minutes and flash rates over this time obtained per luminous
dinoflagellate. In one series of experiments undertaken at 9 °C,
the average flash rate of a *Peridinium* was found to be 0.9 in
15 minutes. These measurements were made in the absence of
deliberate water agitation, but random fluctuations caused by

Figure 7.7 Luminous dinoflagellate display along the Gower Peninsula, South Wales, summer 2023. Reproduced with permission from Clive Davies.

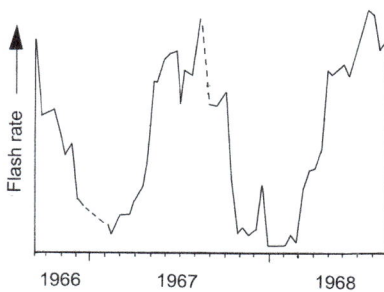

Figure 7.8 Spontaneous flash rate of *Peridinium* and other luminous dinoflagellates over a 2-year period showing seasonality. Flash rate is on a logarithmic scale, broken lines indicate data gaps. Cumbrae, Scotland. Reproduced from ref. 14 with permission from Cambridge University Press, Copyright 1971.

other plankters within the samples may have been responsible as noted above. By regular sampling over a period of nearly 2 years, he was able to demonstrate a strong seasonality of these species with maximum activity occurring in mid-summer (Figure 7.8).

7.6 MARINE BACTERIA

Bacteria are ubiquitous in the oceans and luminous forms appear to be cosmopolitan but they rarely reveal themselves unless in large numbers. They are particularly noticeable on the surface of dead fish and squid and it was J. F. Heller in 1853 who first suggested that a 'Pilz' (microbe) which he named *Sarcina nocti-luca* was the source of this light. Improvements in microscopy and culture techniques revealed that bacteria play an important role in the light organs of some marine fishes. Bacteria often enter into symbiotic relations with other organisms providing benefits to both in the form of protection, nutrition and reproduction. *Noctiluca* for example is sometimes found with apparently healthy bacteria within its cells but whether this is related to its bioluminescence is not known. Planktonic bacteria have been implicated in phenomena known as 'milky seas'.[15] They are distinguished from *Noctiluca* blooms in being more diffuse and not being influenced by water agitation and turbulence. Boats passing through them have not been seen to increase the luminescence in their wake and on occasion have even reduced it. Although there are over 200 documented reports of these occurrences there have been few studies and most of these are from the north-west Indian Ocean and the seas near Java. Milky seas have a duration of several hours to days and at least some are associated with phytoplankton. This has led to the suggestion that luminous bacteria occur within the blooms and are dependent upon them for sustenance. Although planktonic bacteria are common in the sea all require an energy source. Since they are heterotrophs this energy has to come from organic matter which is ultimately the product of photosynthesis. For free-living (*i.e.* non-symbiotic) bacteria virtually all will come from the phytoplankton although a small part will originate from the larger seaweeds and organic matter brought into the sea by streams and rivers. While some bacteria will thrive on the dissolved organic matter of seawater, it is likely that most of them are associated with particulate organic matter because here the food source needed for energy is concentrated in one place. Bacteria have effective mechanisms for attaching to surfaces and it is unusual when examining phytoplankton not to find them on their cell walls or within their mucilage.

In the 1980s a research vessel passed through the Western Arabian Sea into a large area of water that appeared luminescent and had all the characters of a milky sea.[16] Water samples were examined and found to contain a soup of luminescent planktonic organisms including dinoflagellates, ostracods, radiolarians and copepods. Luminous bacteria identified as *Vibrio harveyi* were also found attached to gelatinous colonies of the phytoplankter *Phaeocystis* and the researchers concluded that these bacteria were responsible for most of the luminescence.[17] *Phaeocystis* is well known to marine biologists as it often forms dense blooms especially in the later stages of growth when mucilaginous bodies develop containing thousands of small cells. Some *Vibrio* species also produce an alginase, an enzyme capable of breaking down algal mucilages. A further study employing satellite sensing 17 years later reported a region in the Indian Ocean that was seen to glow for three successive nights.[17] No sampling was possible but the researchers estimated that the amount of light emission would have required the presence of 2.8×10^8 *Vibrio* cells per square centimetre of the sea surface, a huge number. Dinoflagellates were discounted because their light emission would not have been sufficient to have been detected over such a large area although *Noctiluca* is also capable of forming enormous blooms as noted above. Other unusual forms of marine luminescence can be found in early reports from ships plying the oceans and include regular bands and rotating beams of light for which there is no explanation.

The bacterium genus *Vibrio* contains about 20 species, five of which are known to be luminescent.[18] They are mainly found in the sea and virtually all of them have a high sodium requirement for growth making them less tolerant of freshwaters. *V. harveyi* is thought to be the most significant marine member, at least in the North Atlantic and has been isolated from seawater on many occasions. This species, along with most of the others, is a facultative anaerobe, in other words, it respires using oxygen if it is present and in its absence it can continue to make energy and grow by the process of fermentation. This gives the organism a considerable advantage in environments where oxygen levels vary such as those within a dense algal bloom. Here, respiration at night could severely reduce oxygen levels in the water. It has been hypothesised that luminous facultative anaerobes such as *Vibrio* may have evolved from fully anaerobic organisms. Anaerobic microbes would not be able to tolerate oxygen in their cells and a mechanism such as that provided by the luciferin–luciferase system would remove it. The emission of light would be fortuitous and of no advantage to the organism. While this remains an appealing suggestion, there are other biochemical pathways within bacterial cells that can remove oxygen that may in fact be more efficient. This probably bears testament to the fact that most facultative anaerobes are non-luminescent. Dr A. Czyz and colleagues calculated that bacterial luminescence requires a significant amount of energy and discovered that weakly-luminescent mutants of *Vibrio harveyi* could not repair damaged DNA if left in the dark, but were able to do so in the light.[19] This suggested that light emission plays a role in their DNA repair and would be particularly important for the deep-sea bacteria. Whole-genome comparisons of some of these bacteria would be worth undertaking. Unlike dinoflagellates, bacteria have no internal compartmentalisation in which to place their light-giving reactants but the emission, which is continuous, does appear to be localised to some extent within their cells.

Some interesting experiments with cultures of luminescent bacteria have shown that emission can depend upon their cell density. When densities are low, little if any emission is observed, but as it increases, emission also increases dramatically. This has been shown to be due to the synthesis of a molecule which is capable of switching on luciferase synthesis once numbers reach a critical level and the process has been dubbed 'quorum sensing'. It is now widely recognized in other non-luminescent bacterial systems. Whether bacteria attached to cells in a bloom of *Phaeocystis* behave in this way is not known for certain. Numbers of bacteria inhabiting the mucilage of live algae are usually low but would be expected to rise once the bloom begins to decay. This might explain the comparatively short life of milky seas, of the order of a few days as the bloom begins to deteriorate but this is not the whole story. Some luminous bacteria do not appear to respond to quorum sensing and at least one bioluminescent marine bacterium, *Altermonas hanedai* is not facultatively anaerobic but an aerobe.

Other luminous marine bacteria are known. There are two bioluminescent species of *Photobacterium*, *P. phosphoreum* and *P. leiognathi*, common surface-dwellers that are similar in many respects to *Vibrio* but accumulate a refractive substance within their cells. They are also found as symbionts in the luminous organs of marine fish but none of these bacteria can be identified with certainty without isolation and culture on growth media. Luminous bacteria are present in the luminous organs of many marine animals and have been well investigated. The reader is guided to the further reading section for more information. Finally the 'milky seas' formed by the coccolithophorids deserve mention. These belong to another group of microscopic marine algae all of which have the facility to form minute crystals of calcium carbonate on their cell walls. This mineral is very effective at scattering light, and blooms of coccolithophorids, some of which cover enormous areas of the ocean, turn the sea milky during the daylight hours. They are not luminescent and cannot be seen at night but it would be interesting to see the result of mixing a coccolithophorid bloom with a bloom of luminescent bacteria or dinoflagellates.

There are several other groups of planktonic organisms which are capable of illuminating the sea apart from bacteria and dinoflagellates. Some radiolarians, ostracods and shrimps are self-luminous but their emission is generally weak and they are usually only abundant in the warmer seas.

FURTHER READING

M. Anctil, *Luminous Creatures*, McGill-Queen's University Press, Montreal & Kingston, 2018.

C. Castellani and M. Edwards, *Marine Plankton*, Oxford University Press, 2017.

S. H. D. Haddock, M. A. Moline and J. F. Case, Bioluminescence in the Sea, *Ann. Rev. Mar. Sci.*, 2010, **2**, 443–493.

B. M. Sweeney, Bioluminescence and Circadian Rhythms in Dinoflagellates, in *Biology of Dinoflagellates*, ed. F. J. R. Taylor, Blackwell, Oxford, 1987, pp. 269–298.

T. Wilson and H. J. Woodland, *Bioluminescence: Living Lights, Lights for Living*, Harvard University Press, Cambridge, Ma., 2013.

REFERENCES

1. R. Dubois, Note sur la physiologie des Pyrophores, *C. R. Seances Soc. Biol.*, 1885, (8)**2**, 559–562.
2. E. N. Harvey, *A History of Luminescence: From the earliest times until 1900*, American Philosophical Society, Pa., 1957.

3. D. Bottoni, *Pyrologia Topographina id est De igne dissertation juxta loca cum eorum descriptionibus*, 1692.
4. H. Baker, *Employment for the microscope*, London, 3rd edn, 1785.
5. J. Macartney, Observations upon Luminous Animals, *Phil. Trans. R. Soc.*, 1810, **100**, 258–293.
6. A. de Quatrefages, Observations sur les noctiluques, *Ann. Sci. Nat., Zool.*, 1850, (3)**14**, 236–281.
7. C. C. Davis, Concerning the flotation mechanism of *Noctiluca*, *Ecology*, 1953, **34**, 189–192.
8. J. W. Hastings, M. Vergin and R. J. DeSa, Scintillons: the biochemistry of dinoflagellate bioluminescence, in *Bioluminescence in Progress*, ed. F. A. Johnson and Y. Haneda, Princeton University Press, Princeton, 1966, pp. 301–335.
9. R. Eckert and G. T. Reynolds, The subcellular origin of bioluminescence in *Noctiluca miliaris*, *J. Gen. Physiol.*, 1967, **50**, 1429–1458.
10. M. Valiadi and D. Iglesias-Rodriguez, Understanding bioluminescence in Dinoflagellates – How far have we come?, *Microoorganisms*, 2013, **1**, 3–25.
11. J. A. C. Nicol, Observations on luminescence in *Noctiluca*, *J. Mar. Biol. Assoc. U. K.*, 1958, **37**, 535–549.
12. Q. Xie, N. Yan and X. Yang, *et al.*, Synoptic view of an unprecedented red *Noctiluca scintillans* bloom in the Beibu Gulf, China, *Sci. Total Environ.*, 2023, **863**, 160980.
13. J. MacCullough, Remarks on marine luminous animals, *Quart. J. Sci. Litt. Arts*, 1821, **11**, 248–260.
14. P. B. Tett, The relation between dinoflagellates and the bioluminescence of sea water, *J. Mar. Biol. Assoc. U. K.*, 1971, **51**, 183–206.
15. P. Herring and M. Watson, Milky seas: a bioluminescence puzzle, *Mar. Obs.*, 1993, **63**, 22–30.
16. D. Lapota, C. Galt, J. R. Losee, H. D. Huddell, J. K. Orzech and K. H. Nealson, Observation and measurements of planktonic bioluminescence in and around a milky sea, *J. Exp. Mar. Biol.*, 1988, **119**, 55–81.
17. S. D. Miller, S. H. D. Haddock, C. D. Elvidge and T. F. Lee, Detection of bioluminescent milky sea from space, *Proc. Nat. Acad. Sci. U. S. A.*, 2005, **102**, 14181–14184.

18. J. Poupin, A. Cussatlegras and P. Geistdoerfer, *Plancton marin bioluminescent*, Rapport Scientifique du Loen, Ecole Navale, Laboratoire d'Océanographie, France, 1999.
19. K. C. Mok, N. S. Wingreen and B. L. Bassier, Vibrio harveyi quorum sensing: a coincidence detector for two auto-inducers controls gene expression, *Embo J.*, 2003, **22**, 870–881.

CHAPTER 8

Luminous Land Plants, Microbes and Fungi

8.1 INTRODUCTION

The terrestrial vegetation although of great diversity, ranging from microscopic algae to some of the largest organisms on earth is not noted for its luminescence. There are nevertheless a few intriguing observations regarding the luminescence of terrestrial plants, some of which still await confirmation. The illumination of foliage by St Elmo's Fire has already been noted and other electrical phenomena have occasionally surfaced as in the discharge of pollen from flowers. Thomas Phipson reported light emission from the breaking of the spathe which envelops the *Pandanus* flower.[1] It was accompanied by a sharp cracking noise. However similar observations obtained with arum have been discounted.[2]

The luciferin–luciferase reaction with its glowing intermediates has never been observed in terrestrial green plants, at least until recently when geneticists managed to insert genes into some crop plants causing them to glow. Other cases of bioluminescence have since been shown to be false being caused by novel forms of light reflection as in the case of the moss *Schistostega*.

Luminous Phenomena: A Story of Spontaneous Combustion, Phosphorescence and Other Cold Lights
By Allan Pentecost
© Allan Pentecost 2025
Published by the Royal Society of Chemistry, www.rsc.org

Among the non-photosynthetic terrestrial plants however there are a good number of well documented cases. These are from the fungi and bacteria along with some of their animal associates. Bacterial light has been reported mainly from decaying animal remains and in the fungi. A small number of species have been found to be effective emitters.

8.2 VASCULAR PLANTS AND MOSSES

There are several reports of luminescence in flowering plants, particularly red and orange coloured flowers such as those of the marigold, poppy and nasturtium. Observers have reported a fleeting emission although doubts have been expressed over its existence. Flower luminescence could have evolved along with insect pollination, since some insects are attracted to flower colour but it is evidently not a common phenomenon if it exists at all. It deserves further investigation. One suggestion is the static electric discharge from pollen grains but this does not appear to be likely. Some flowers have often been seen to glow in near darkness and at this time the relative proportion of near-UV to sunlight is increased and some flowers are known to have fluorescent petals, a clear adaptation to insect attraction. Another as yet unsubstantiated occurrence of luminescence in plants is in the milky latex of *Euphorbia phosphorea*. A native of the Brazilian dry tropics, the latex was reported by its discoverer, von Martius, to glow in the dark after being brushed on a surface, particularly upon warming. *Euphorbia* latex contains a huge range of substances in which terpenoids tend to predominate. These are complex biomolecules consisting of a terpene framework which carries one or more oxygen-containing functional groups. There is currently no evidence to support luminescence in plant latex but with such a diversity of compounds occurring in *Euphorbia* species it remains a possibility. If correct it is unlikely to be used to attract insect pollinators or deter grazers as the latex is only visible when the plant is damaged.

Advances in gene technology have led to the production of 'glow in the dark plants'. Originally novelties they have proved valuable in selecting new varieties for agriculture and biotechnology. The first plant to be engineered in this way was the tobacco plant. DNA coding for firefly luciferin was put into a bacterium species and the bacterium was then used to transfer the gene to the plant. In the soybean, bacterial genes for luciferase were placed next to the promoters for the enzyme nitrogenase, an important enzyme for the fixation of nitrogen. Measuring the amount of luciferase in these plants gives an estimate of nitrogenase activity. This can be done rapidly allowing the farmer to determine how much nitrogen fertiliser would be needed to feed the crop. Since then several other methods have been made to make plants glow. One uses extremely small particles of silica with sorbed emitting components. Using high pressure the particles find their way into the plant leaves. Those containing luciferase enter the plant cells while those containing luciferin are placed within the extracellular space which includes the cell walls. The luciferase then diffuses slowly into the cells making them glow for several hours. Recently, strains of the bioluminescent and UV-fluorescent orchid *Phalaenopsis* have been developed. Similar effects have been obtained by feeding plants with artificial phosphors (see Chapter 6).

The moss *Schistostega pennata* is remarkable for its glowing protonema.[3] The protonema is an early vegetative stage of a moss formed by germinating spores. In most mosses it is short-lived and consists of fine branched filaments the cells of which contain chloroplasts making it look like a green alga. *Schistostega* is unusual on three accounts, first its protonema consists of aggregates of spherical cells, second the protonema is perennial and third, it is a species of sandy caves where it may be seen growing on the walls and floors. When a point source of light illuminates the cells, such as sunlight at a cave entrance or a torch, the cells glow a brilliant green giving rise to its popular name of 'goblin gold' (Figure 8.1). Light is partially focussed by the cells onto chloroplasts situated at the far end of the cells (Figure 8.2). Most of the red and blue light is absorbed by the chloroplasts but the green light is scattered back towards the light source (Figure 8.3). Since almost all of the returning light is green, rainbow colours cannot be seen in *Schistostega* and in any case the cells would be too small to show good colours owing to interference effects. In order for the light rays to become refracted at the cell surface, it must be reasonably dry and the cells may have a water-repellent surface. The reflected light is reminiscent of the 'cats eyes' placed on roads to improve night

Figure 8.1 *Schistostega* protonema growing in a small cave on Alderley Edge, Cheshire.

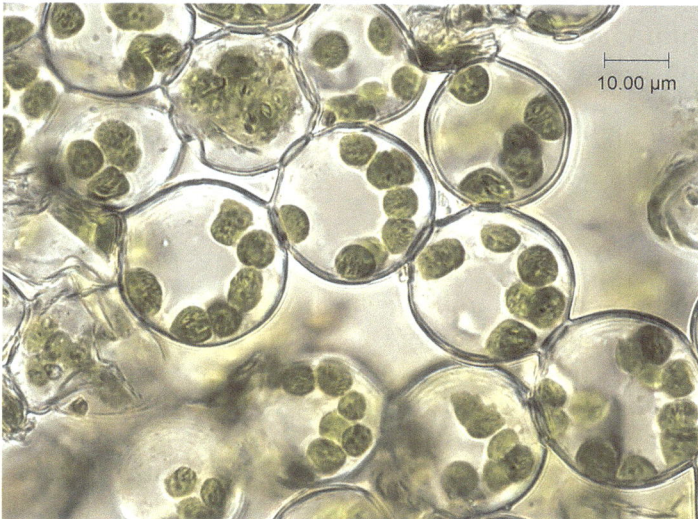

Figure 8.2 Micrograph of *Schistostega* protonema showing the spherical cells containing green chloroplasts. In some of the cells the chloroplasts are arranged favourably for light reception in a cave environment. Image by Chris Carter.

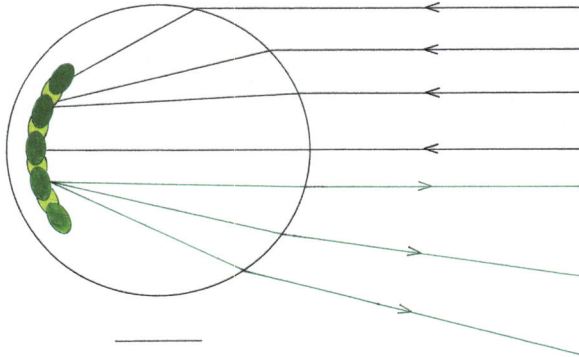

Figure 8.3   Diagram of a spherical protonema cell of *Schistostega* showing four incoming light rays from a distant source, refracted at the cell wall and partially absorbed by the chloroplasts. Three green rays scattered from one of the illuminated chloroplasts are shown below. Bar 10 μm.

driving although in this case a mirror is placed behind a glass lens to optimise the reflection. Nevertheless the illumination provided by this moss is spectacular and there may be other factors at work, perhaps related to the shape and orientation of the chloroplasts. In some plants the chloroplasts are capable of moving to intercept the maximum amount of light, so small differences in the shape of the cells are probably unimportant. Another moss, *Mittenia* also shows this effect and occurs in the Southern Hemisphere.

8.3   FRESHWATER AND TERRESTRIAL ALGAE

No freshwater algae are known to be bioluminescent but two species have been described as being luminous under certain conditions. The first is the common semi-terrestrial and aquatic diatom *Melosira varians*. Blooms of this diatom in Lough Neagh, Northern Ireland, have been described as showing a faint luminosity during the day but there is no evidence that the alga contains a luciferin–luciferase system. The cell walls of diatoms contain silica whose refractive index is slightly higher than that of water but the resulting light refraction would not be sufficient to focus light onto the chloroplasts. The glow may result from light reflection from the cell surface. *Melosira* however has also

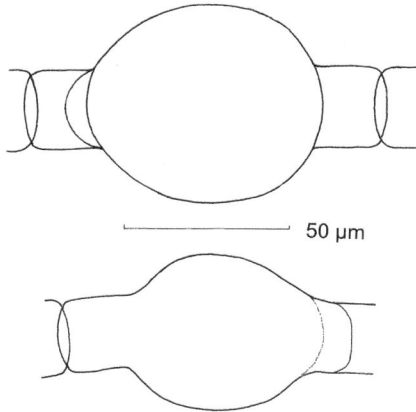

Figure 8.4   Two auxospores of *Melosira varians* drawn from a microscope image. Auxospores are part of the sexual reproduction stage of diatoms. The mature spore walls consist of large numbers of disc-like scales of silicon dioxide embedded in an organic matrix. In section the spores are circular.

been found in caves and there are several reports of cave walls being illuminated by this species. *Melosira* often produces sub-spherical auxospores (Figure 8.4) and it may be that these cells reflect light in a similar manner to the protonema of *Schistostega*. This would require the auxospore surfaces to be exposed to the air. Alternatively, the cells may orientate themselves towards the light source in such a way that the silica walls behave as a re-flector. The phenomenon was last observed in Britain during WW2 when the diatomist John Carter visited a tin mine in north Cornwall, discovering them close to the entrance.[4] The second alga has the unusual property of sitting on the water surface when producing its cysts. In this chrysophyte alga, known as *Chromophyton rosanoffii* the cysts are roughly spherical and si-licified. The cysts contain chloroplasts and are effective at re-turning the sun's rays in a manner similar to that of *Schistostega* and impart a bright glow to the water surface. It has been seen recently in the London Natural History Museum garden pond.[5]

8.4   FUNGI AND LICHENS

Luminescent fungi are more widely reported than terrestrial bacteria with some of the species common and widespread[6]

Figure 8.5   Bioluminescent fungi (*Mycena* sp.) growing on rotting wood.
© Anna Pluciniska, Shutterstock.

(Figure 8.5). In northern Europe most of this luminescence is caused by the Honey fungus *Armillaria mellea*. The common luminous species in southern Europe is *Omphalotus olearius* and in North America the related *O. illudens* where it is sometimes termed the 'Jack o' Lantern'. Both have luminous fruit bodies and are also poisonous. In all over 100 species of fungi are known to be bioluminescent and in most cases it is usually either the mycelium or fruit body that is involved although the emission is often weak. In a few species only the gills luminesce and in others it is both mycelium and fruit body. In the tropical *Roridomyces roridus* luminescence is only seen in the spores. The fungi are an extremely diverse group of organisms but only one bioluminescent species has been found outside the familiar toadstool group, the Basidiomycota. This is the ascomycete *Xylaria hypoxylon* also known as the candle snuff fungus. It is common throughout Britain and much of Europe. Light emission in the fruit body of this fungus is so faint that amplification is required using a long photographic exposure or a photomultiplier. Since many organisms show low levels of emission there are probably other ascomycetes that are weakly luminescent. Emission in the fungi has been the subject of much

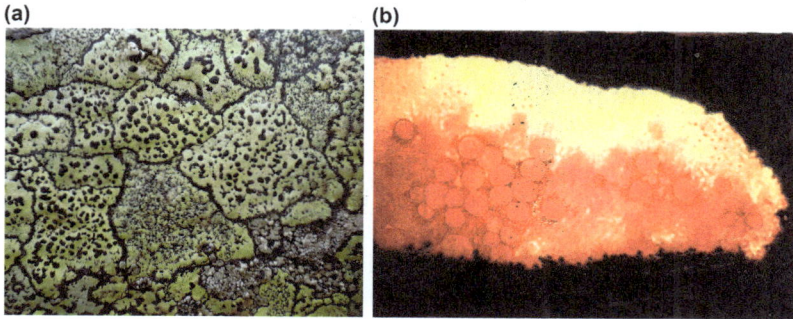

Figure 8.6 (a) The 'map lichen' *Rhizocarpon geographicum* forms luminous yellow-green patches on rocks worldwide. Image 6 cm across. (b) A section through *Rhizocarpon* illuminated with UV radiation showing the distribution of rhizocarpic acid which fluoresces yellow. Below this, chloroplasts of the symbiotic green algae fluoresce red. Section about 0.3 mm thick. Reproduced with permission from Steve Clayden.

speculation as to its purpose. One suggestion is that the light given off by the Honey fungus is used to attract night-flying insects in order to help disperse the spores but this has not been widely accepted. All of the luminous species appear to be 'white rot' fungi feeding off decaying wood and it has been suggested that the emission provides protection against oxygen damage during wood degradation. However, the majority of white rot fungi are either non-luminous or only very weakly luminous.

Lichens are fungi that have a close association with one or more algae. Many occur in surprisingly harsh environments such as deserts and mountain tops where they are often revealed by their brilliant colours. One of the most familiar of the mountain species is *Rhizocarpon geographicum*, often called the 'map lichen' (Figure 8.6a). This species forms bright yellow-green crusts often surrounded by a dark line and is found throughout the world providing there is sufficient rainfall. The yellow-green is produced by the fungus and is one of the many 'lichen substances' unique to this group of organisms.[7] It is called rhizocarpic acid and it fluoresces strongly in the yellow region of the spectrum under UV. A thin section through this lichen shows that the acid occurs at the lichen surface (Figure 8.6b). Ultraviolet radiation is damaging to living cells and organisms that grow slowly and live at high altitudes are particularly prone to this. Rhizocarpic acid absorbs much of this radiation and some of the light that is re-emitted as fluorescence finds its way to the algae below and used in photosynthesis.

REFERENCES

1. T. L. Phipson, *Phosphorescence*, Lovell Reeve, London, 1862.
2. E. N. Harvey, *A History of Luminescence: From the earliest times until 1900*, American Philosophical Society, Pa., 1957.
3. D. H. S. Richardson, *The Biology of Mosses*, Blackwell, Oxford, 1981.
4. J. R. Carter, Bioluminescence in *Melosira varians*, *Int. Rev. Speleol.*, 1967, **2**, 407–408.
5. *The freshwater algal flora of the British Isles*, ed. D. M. John, B. A. Whitton and A. J. Brook, Cambridge University Press, 2nd edn, 2011.
6. J. Ramsbottom, *Mushrooms and Toadstools, a Study of the Activities of Fungi*, New Naturalist 7, Collins, London, 1953.
7. R. Lücking and T. Spribille, *The Lives of Lichens*, Princeton University Press, 2024.

CHAPTER 9

The *Ignis Fatuus* or Will o' the Wisp

9.1 INTRODUCTION

In the dank hollows of the countryside, where the land is often neglected owing to the poor quality of its soil, may be found patches of waterlogged ground that rarely dry out, even in the summer sun. Come the dusk of autumn, thin mists gather as the cooling air descends into the humid windless valleys. The light fades and the night silence is broken only by the call of the nightingale. There is a faint rumble accompanied by the steady beat of horses' hooves, as a night coach descends a dirt road into the valley. The night is cold and the coach curtains are closed, for there is little to see, but a curious traveller parts them to reveal a sombre scene. The blackness of the ground contrasts little with the indigo sky above. Without warning, a flash of light is seen but the traveller, unprepared, dismisses it as a reflection from the oil lamp within. With renewed curiosity the traveller's eyes return to the window. There is another flash, then a flicker of light nearby followed by yet another. The ground is alight with fire. But this is no ordinary fire, it is the *ignis fatuus* or 'Will o' the wisp'.

The above fictitious account would have been almost commonplace in parts of England up to the end of the 18th century.

Luminous Phenomena: A Story of Spontaneous Combustion, Phosphorescence and Other Cold Lights
By Allan Pentecost
© Allan Pentecost 2025
Published by the Royal Society of Chemistry, www.rsc.org

Descriptions of lights appearing suddenly at night in damp valleys abound and can be traced back to early times. The unpredictability of the phenomenon and difficulty encountered by would-be investigators in coming face-to-face with these lights has given them a supernatural image and made them the subject of numerous myths and legends. Their association with bogs and marshland, and their nocturnal character add further hazards for the field researcher. Despite these difficulties, several intriguing and close-hand accounts are available. They help build a picture of the phenomenon but are insufficiently detailed to permit a reliable explanation. Several accounts, while seemingly reliable, contain important but conflicting details, leading one to suspect that several processes are involved. 'Will o' the wisp' is normally only encountered nowadays as a figure of speech indicating something that is fleeting or out of reach. The phenomenon is rarely reported nowadays and possible reasons for this will be explored later.

This chapter begins with an examination of the folklore associated with these lights. Accounts will be given from several parts of the world and some common strands will be pulled together. There follow a series of first-hand descriptions of close encounters accompanied by a map illustrating their distribution in the United Kingdom. There can be no guarantee that all accounts are genuine, but again there are some common themes, this time set against a few apparently irreconcilable problems. They are intriguing tales. Possible explanations are then examined. These range from the chemistry of some unstable gaseous compounds believed to be formed by bacteria to forms of biological luminescence. The origin of these gases in the natural environment is examined and the section is concluded with some microbiological phenomena that may account for them. The phenomena are sometimes so faint as to border on the limits of visibility. Finally the apparent demise of the *ignis fatuus* is discussed along with some of the author's own attempts at their investigation.

9.2 FOLKLORE

Radio broadcasts first became widely available in the United States and northern Europe in the 1920s when they rapidly achieved popularity and began to supplant other forms of entertainment. Prior to this, group activities within the home mainly resided in storytelling and music-making. Stories were often read aloud from books or from memory, with fables and legends a popular choice among all walks of life. Folklore encompasses traditional customs and beliefs and these too will be discussed. Before 1800, so little was understood about nature, that beliefs considered outrageous today were commonplace and accepted without question. It became apparent that most folk tales originated from rural areas where traditions were stronger than in the cities and contradictory phenomena such as 'fire without heat' were topics ripe for incorporation into folklore. Such is the case with the 'foolish fire' or *ignis fatuus* as it was often called.

Four themes feature strongly in folklore: aspects of nature, warning of danger, religion and death – and all have some connection with waterbodies. Water is both revered and reviled. Clean waters have always been held in high regard and were of paramount importance prior to the advent of water distribution systems and disinfection. Holy wells remain places of pilgrimage and worship in most areas of human habitation. The recent rise in the bottled spring water industry attests to the enduring powers of certain waters and there remain many springs connected historically with healing and convalescence where people continue to bathe. Water is also feared. When still, the reflection of light from a lake surface prevents its depth from being seen. When disturbed, waves, bubbles, foam and suspended mud further reduce the visibility. Where the bottom can be seen, it is usually dark and its depth is apparently reduced by the refraction of light. Combine these uncertainties with the darkness of a starlit night and the potion becomes a powerful mix with which to spin tales of fear and the supernatural. Below is a selection of myths connected with the *ignis fatuus* aligned with three of these folklore themes: warnings of danger and deception, religion and death.

9.2.1 Warnings of Danger and Deception

The Register of Deaths for Lamplugh, Cumbria, from January 1, 1658, to January 1, 1663:

> *Fright to death by fairies 6; bewitched 4; Old women drowned on trial for Witchcraft 3; led into a horsepond by a Will o' the Wisp 1.*

This is a common theme into which many stories belong, which appears to have begun some time in the Middle Ages. Further examples are given below.

The wherryman

Up to the mid-19th century, the River Yare in Norfolk was an important passageway for the transportation of goods between the port of Great Yarmouth and Norwich. Small wooden sailing boats called wherries were used and the wherrymen would moor their boats at riverside staithes. These men would congregate at the White Horse Inn at Thurlton where the marshes surrounding the river extended for a mile or more in width. One evening, a ferryman, whose home was in Thurlton, decided to return to Norwich on his boat to collect some provisions for his wife.[1] Night sailing was unusual on the Yare and he was warned by one of the company- '*Anyone trying to cross the marshes to the staithe at night is daft – I have been a long time finding my way here. It is pitch dark and the Jack o' Lanterns are popping off in hundreds. I lost my way twice. The first time I was stopped by a dried-up dyke; the second time there seemed an untold number of Jack o' Lanterns floating about. I was mighty glad when I saw the lights shining from this pub's windows*'. The advice was not heeded by the Thurlton wherryman. '*Goodnight to you all, I will be seeing you in the morning; I'm not worried about those Will of the Wisps; I know the old marsh too well for them to lead me astray*'. The remaining wherrymen spent the night at the inn, and at daybreak made their way back to their wherries. The Thurlton man was not to be seen. Concern for the man's fate grew a few days later when a body was seen floating on the tide toward Norwich. It returned on the ebb tide nearly to Yarmouth and it was then seen in the River Bure. The wherrymen would not touch it out of superstition and finally it was washed up between Reedham and Breydon. The Thurlton wherryman was buried in the village churchyard where his tomb is said to still be seen.

Another version of this story was published by *Countryfile Magazine*. Here the man is named Joseph Bexfield and the date is given as 11 August 1809. Joseph is found at the inn looking over the marshes toward the staithe, where a small scurrying figure is seen with a lantern moving rapidly to and fro. The wherryman fears for the safety of his boat and plans to investigate. He is warned of the Lantern Men, who may steal the breath from your lungs: '*If a lantern man is upon ye, throw yeself flat on ye face and halt ye breathing*' says one. Undeterred, Joseph heads off toward the light and is found dead 3 days later by the staithe. His gravestone can be seen at All Saints Church, Thurlton.

The first storyteller goes on to recall an evening service on the marshes. The parson repeated '*I will be a light unto your feet*' and the congregation proceeded to sing:

Lead kindly light amid the encircling gloom
Lead thou me on
The night is dark and I am far from home
Lead thou me on

On leaving the church they were confronted with hundreds of flickering lights on the marsh.

The story of Pwca

This tale comes from Wales and was related by William W. Sikes in his *British Goblins*.[2] Sikes was an American writer who was appointed later in life as United States Consul in Cardiff. He became interested in Welsh folklore and customs and gathered data from earlier published sources as well as from original oral traditions.

A peasant is returning home from the fair when he sees a light travelling before him. Looking closer he sees that it is carried by a dusky little figure, Pwca holding a lantern or candle at arm's length over its head. He follows it for several miles, and suddenly finds himself on the brink of a precipice. From far below, there rises the sound of a foaming torrent. At the same moment the little goblin with the lantern springs across the chasm, alighting on the opposite side; raises the light again high over its head, utters a loud and malicious laugh, blows out its candle and disappears up the opposite hill, leaving the awe-struck observer to get home the best he can. Pwca, also known as Pooka may be the mischievous Puck of Shakespeare's Mid-summer Night's Dream:

9.2.2 Religion and Death

The flames of Satan's fires might be expected to correlate with the *ignis fatuus* but there are no stories in the Bible or Quran that appear to relate to them. However these works originated in arid regions where the phenomenon would not have been widespread.

In Lancashire it was said that if the *ignis fatuus* is seen, travellers are advised to thrust the blade of a clasp knife into the ground then lay prostrate upon it. A possible connection between the devil and these phenomena is found in George C. Davies' influential work on the Norfolk Broads.[3] Davies spent much of his spare time sailing in the Broads, where he came across frequent references to the devil in place names. The Broads have long been known for reports of Will o' the wisp.

There was also widespread belief in corpse lights. These are lights occasionally seen in graveyards where there have been recent burials. However, other 'night lights' must have been confused with them on occasion. One can imagine the famous church light at Weldon, Northampton being mistaken for one, although this was designed to lead people by a *safe* path to their destination.

Further afield, The Thonga Tribe of South Africa called the *ignis fatuus* 'witch fire'. It was interpreted as a message sent by witches to terrify wrong-doers. If one was seen by a group of onlookers, measures were taken to extract a confession from an individual to reveal the misdemeanour.

In Lower Saxony and parts of Ireland the *ignis fatuus* is the soul of someone condemned to wander for having moved or disregarded a boundary marker, or a wandering soul of one refused heaven and hell. In Germany it can be the soul of an unbaptized infant. The Irrlicht as it is known in Germany is a general omen of death or a forest spirit. In Irish and Northern English folklore it is a presage of death and in Hampshire the light is said to go out when the soul of the dying departs.

The Penobscot Indians of New England and eastern Canada were hunters and fishermen. They lived along the Penobscot River that flows south into the Atlantic Ocean, north of Portland in Maine. They have many legends about the forests and rivers and the mythical creatures that inhabit them. Some of the stories were told to children as a warning of the dangers that await them in those remote areas. The fire creature of the Penobscot Indians is an omen of death and could be the *ignis fatuus*: but there is nothing in the stories to confirm it.

9.3   LITERATURE, ART, MUSIC AND FILM

9.3.1   Literature and Art

Beowulf is a heroic Germanic poem thought to have been written in Northumbria in the early 8th century. Lines 1365-7 relate that ...*There may be seen each night a fearful wonder – 'fire on the flood'*... This is part of the description of the Mere, a dark lake and abode of the evil monster Grendel and his mother, both of whom Beowulf eventually slays. This could be a reference to marsh lights although it has also been attributed to luminous fungi.

Perhaps the earliest British text mentioning the *ignis fatuus* is in a poem by Dafydd ap Gwilym dating to 1340 where '*There was in every hollow ; a hundred wrymouthed wisps*'. Dafydd ap Gwilym, born near Aberystwyth, is a celebrated Welsh poet. He travelled widely in the principality, and in doing so gained much local knowledge, as is evident from his literary achievements. A little later (*ca.* 1360) William Langland wrote one of the earliest known poems in Middle English called *Piers Plowman*. The subject, Will, falls asleep and Piers appears in a dream as a humble plowman recounting through a series of stories, the virtue of Christianity. It is set close to the Severn Valley where there have been many sightings of Jack o' Lantern in the past. In *Passus* 5, the churlish, ale-drinking, Sir Gloton appears:

> *Til Gloton hadde yglubed a gallon and a gilk*
> *His gutless bigonne to gothelen as two greedy sowes*
> *He pissed a potel in a Paternoster while,*
> *And blew his rounde ruwet as his ruggebones ende*
> *That alle that herde that horn helde hir noses after*
> *And wished it hadde ben wexed with a wisp of firses!*

Allusions are also made to the *ignis fatuus* in Michael Drayton's poem *Nymphydia* (1627) where King Oberon's Queen Mab, seduced by Pigwiggen, engages Puck to search the couple out. Of Puck (also known as Hob) we hear:

Of purpose to deceive us
And leaving us makes us stray
Long winter's nights, out of the way
And when we stuck in mire and clay
Hob doth with laughter leave us.

In Edmund Gayton's *Festivous Notes upon Don Quixote* (1654), in discussing the actions of the Don he notes: '*Witches are confin'd in their night rambles to egge shels, and Hell affords nothing but an Ignis fatuus, an exhalation, and Gillion a burnt taile, or Will with the wispe*'.

The phenomenon is also mentioned by many poets including Samuel Butler, Robert Burns, Edward Young, Tom Parnell and Thomas Gray. Some of these were active in the 18th century and the last three are known as the 'graveyard poets' celebrated for their gloomy works. Believed to have been inspired by Milton's *Paradise Lost*, they were the forerunners of the gothic English writers of the next century such as Mary Shelley and Bram Stoker.

References to the *ignis fatuus* are common in some of the classical novels of the 19th century. Charlotte Brontë refers to it on five separate occasions in her novel *Jane Eyre*. Living at Haworth Parsonage among the bleak Pennine moors, the Brontes may well have observed the phenomenon but there is no mention of it in their private writings.

Will o' the wisp is not a popular subject for artists and there are no well-known works on display in public galleries. But there are a good number of book illustrations containing woodcuts and engravings from English, French and German works. One of the earliest is an engraving in *Meteorologia* by F. Reinzer, first published in 1697 with the illustration by J. Kadoriza appearing in 1709. This is described as '*an atmospheric ghost light seen by travellers at night over bogs, like a flickering lamp*'. The work shows a small group of individuals holding what appear to be flaming tapers running across hills and fields at night with little reference to marsh lights. Jabez Allies in his book *Will o' the Wisp and the Fairies*[4] mentioned a picture exhibited at the Royal Academy with a horse rider holding an *ignis fatuus* in his hand. A similar theme is found illustrating an edition of Robert Burn's poems dated around 1800. Here a mischievous imp holds a burning lamp above an open marsh. A coloured lithograph by an unknown artist appeared around 1820 showing a flying demon holding a bright candle to the astonishment of a traveller walking through a mire. Later, the English artist Arthur Rackham painted a watercolour of Puck holding a shining light above an open marsh in an early 20th century edition of Shakespeare's *A Midsummer Night's Dream*.

What would appear to be more realistic depictions of the phenomena can also be found. In a work dating to 1750 in Bayard's *Eclaires tonnerres* the Abbé Richard is seen in a grave-yard at night astounded by 4-metre high flames rising from open ground. Another graveyard scene appears in *L'Atmosphere* by Camille Flammarion (1871). Here the anonymous artist shows flames about 30 cm high arising from marshy ground. They burst over recent graves of communards executed after the Commune at Issy, near Paris (Figure 9.1a).

In the 1811 edition of Casell's *World of Wonders* is a significant illustration that claims to show an actual sighting in Lincoln-shire by the illustrator William Pether (Figure 9.1b). The light appears about 2 m above a wooded marsh and is approximately 40 cm high. One of the best examples is a hand-coloured en-graving by the English artist J. W. Whymper appearing in a series of plates entitled *Natural Phenomena* first published in 1846. In it a lone traveller is seen walking towards a reedy morass on a starlit night. Hidden among the reeds and about 30 cm above the water a brilliant light throws a strong reflection on the water surface (Figure 9.1c). Another interesting work is an oil painting by C. Spitzweg *ca.* 1870 entitled *Das Irrlicht* held in the Bavarian State Picture Gallery. It shows a group of faint lights rising from a pool in open countryside. A further example is found in an en-graving showing a small flame rising from shallow water in the 1869 edition of *Les Météores* by Margollé (Figure 9.1d). In these later works the lights appear to be 20–40 cm in height rising directly from the water.

9.3.2 Music and Film

Marsh lights are referred to in the second of the song-cycles composed by Franz Schubert known as the *Winterreise*. It is a

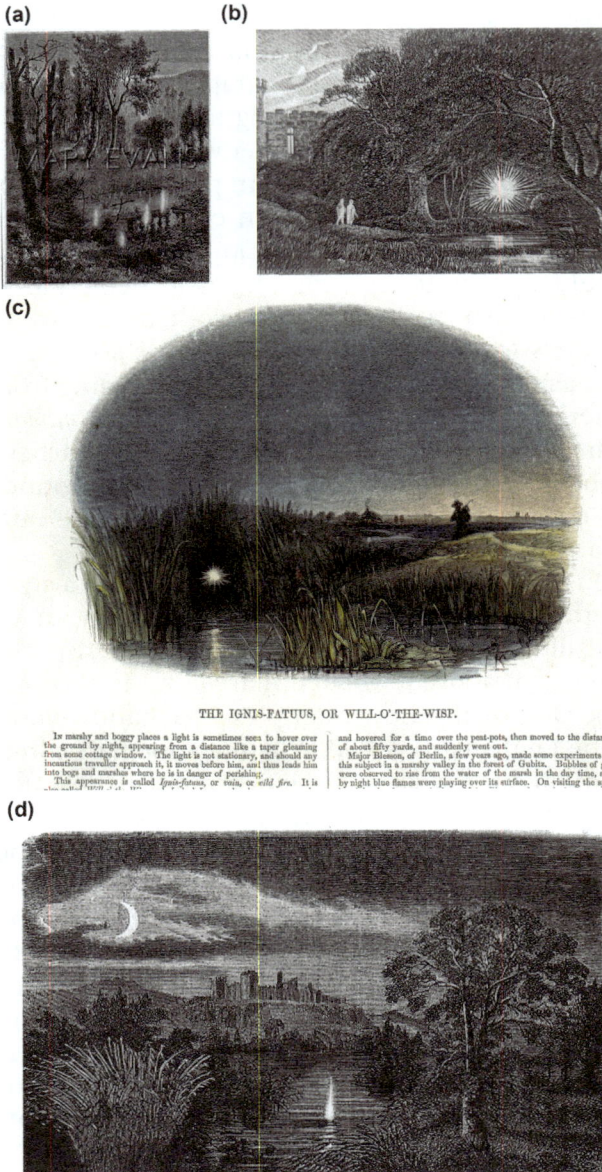

(a)

(b)

(c)

THE IGNIS-FATUUS, OR WILL-O'-THE-WISP.

In marshy and boggy places a light is sometimes seen to hover over the ground by night, appearing from a distance like a taper gleaming from some cottage window. The light is not stationary, and should any incautious traveller approach it, it moves before him, and thus leads him into bogs and marshes where he is in danger of perishing. This appearance is called *Ignis-fatuus*, or *rein*, or *wild fire*. It is

and hovered for a time over the peat-pots, then moved to the distance of about fifty yards, and suddenly went out. Major Blesson, of Berlin, a few years ago, made some experiments on this subject in a marshy valley in the forest of Gubitz. Bubbles of gas were observed to rise from the water of the marsh in the day time, and by night blue flame were playing over its surface. On visiting the spot

(d)

Figure 9.1 Early depictions of the Will o' the wisp. (a) The *ignis fatuus* bursting over graves of members of the Paris Commune. From *L'Atmosphere* (1871). (b) An illustration of the Will o' the wisp in Casell's *World of Wonders* (1811). (c) Ilustration by J. W. Whymper in *Natural Phenomena* (1846). (d) A *feux follet* rising from shallow water in Margollé's *Les Météores* (1869).

setting of 24 poems by the German poet Wilhelm Müller and is a story of lost love. A heartbroken young man wanders out of town into the darkness of night searching for solace. In the ninth song entitled *Irrlicht 'the false light of the irrlicht has led him astray, but he is used to that. Every stream reaches the sea, every sorrow its grave'*. Although the *Winterreise* is acclaimed as a major work, it is a dark and gloomy piece. Schubert was known to be ill and depressed during its composition.

Eighteen years later, the French composer Hector Berlioz composed a large choral work, *La Damnation de Faust*. Here, Mephistopheles, the devil disguised as a man, approaches Faust who is lost in despair. Mephistopheles presents him with a young woman, Marguerite with whom he falls in love and is ultimately the cause of his downfall. There follows the *Minuet des follets* (Minuet of the wisps) where the wisps dance in bizarre formations around Marguerite's house.

The Spanish composer Manuel de Falla, in his ballet *El amor brujo* (*The Bewitched Love*) had Candela, an Andalusian gypsy girl dancing every night with her husband's ghost. He had been murdered by his mistress and in an attempt to exorcise her from the ghost the heroine is seen dancing with a Will o' the wisp to mysterious music.

In the *Scarlet Claw*, a 1944 Universal Studios film based upon Conan Doyle's Sherlock Holmes and Dr Watson, Holmes goes out to investigate mysterious goings-on near the village of La Mort Rouge. There he encounters a strange light darting over the marshes and aiming his revolver, Holmes lets off several shots at the apparition. Having been informed of the event Watson enters the scene and is rudely pushed into the bog by a retreating figure. Holmes, after saving Watson detaches a piece of glowing clothing that has become snagged by a tree. After examining the cloth, Holmes declares that it has been impregnated by phosphorus and worn by a mortal. The *Scarlet Claw* was not a work by Doyle but an adaptation including elements of the *Hound of the Baskervilles*. The latter, set in the wilds of Dartmoor, makes no direct mention of marsh lights despite its lurid description of the fictional Grimpton Mire. Here creatures two-legged and four disappeared from time to time into the morass. The location is believed to have been based upon Foxtor Mire where there have been sightings of marsh lights.

9.4 ETYMOLOGY

The phrase Will o' the wisp according to the Oxford English Dictionary was first used in print by John Day in his *Law-trickes* (1608). It is pre-dated by the Latin term *ignis fatuus* (foolish fire) by many years. There are in excess of 60 words or phrases for the phenomenon. Some examples in England are given by J. Allies. He lived in Worcestershire where the lights were well known, particularly in the Vale of Evesham, a low-lying area underlain by clays and shales. Here many local names contain the word *hob*, some of which seem to represent localities where the marsh lights have been seen. Hob is derived from the Gothic word for a horse, and Allies made the tenuous but nonetheless suggestive connection between hob and movements of phosphorescent gas as likened to a cantering horse. In this district he identified several sites with possible connections. They included the Jack Meadows perhaps derived from *Jack o' Lantern*. In the parish of Badsey, the lights were known as Pinket, which may originate from the Dutch verb *pinken*, to wink. He also drew attention to Pink's Field and Pink's Meadow at Dymock near Evesham.

There are at least eight names referring to candles or lamps implying the existence of brightly burning lights rather than a faint glow. *Wisp* is a term for a small twist of straw and *Kit with the Canstick* refers to a lighted candle. In Warwickshire the term *mab-led* signifies being led astray by a Will o' the wisp. This presumably originated from the Mab of fairy lore, reference to which has already been made. Of the thirty or so of the known UK names, many are parochial, being confined to small geographic regions. Examples include the *Hobby Lamp* of Norfolk and the *Solas Uibhist* of the Hebrides. A small number relate to specific locations such as the *Syleham Lamps* of Suffolk. 'Wisp' has at least four variants and can be found through much of England, while separate names usually apply in parts of Scotland and Wales. References to graveyards occur in several names and there are well-authenticated accounts of lights occurring in these locations. In parts of Wales, the *ignis fatuus* was known as the *Ellylldan* and William Sikes observed that ... *Like all goblins of this class, the Ellylldan was seen dancing about in marshy grounds, into which it led the belated wanderer; but as a distinguished resident in Wales has wittily said, the poor elf is now starved to death, and his breath is taken from him: his light is quenched for ever by the improving farmer, who has drained the bog; and instead of the rank, decaying vegetation of the autumn . . . crops of corn and potatoes are grown.*

The *fetch candle* relates to a belief in some areas of the country that a person about to die is seen as an apparition to friends and relatives. Thus these lights are interpreted as an omen of imminent death of an individual. It is known as a *swarth* in Cumbria. The 'wat' in *Joan in the Wat* of northern England may refer to prison-lore where a small light would appear to felons before their appearance at the assizes, signalling their doom.

The earliest known name, and indeed description of the *ignis fatuus* comes from a Chinese poem. It was written during the Western Han Dynasty (206 BCE–24 CE). At least other ten other poems spanning the period 618–1911 CE (Tang to Qing Dynasties) mention it. Further names occur across the world but no comprehensive study appears to have been made. Some examples, mainly from Europe are shown in Table 9.1. A few have no equivalent English translation. One surprising feature of these names is the apparent lack of any early European literature on the subject. For example the Roman author Pliny the Elder undertook a comprehensive survey of the natural world in ten volumes (*ca.* CE 77). He mentioned lightning and other physical phenomena but not apparently marsh lights which must have been known in his time. There is however Dea Marica, sometimes known as the Roman goddess of swamps and a sorceress much feared by the local inhabitants. She was reputed to turn travellers into animals, but even here there is no connection with marsh lights. The Indian *Chir Batti* of the seasonal Banni wetlands have been known for many years. They take several forms, not all of which conform to marsh lights but the connection with wetland is compelling. Marsh lights of Bangladesh known as *Aleya* have superstitions similar to those of Europe. They are thought to represent the ghosts of fishermen who died while at their craft.

9.5 FACTUAL ACCOUNTS

The following accounts provide further insights into the phenomenon.

Table 9.1   Some *ignis fatuus* names in other countries.

Name	Country/region	Notes	Ref.
Buchelmännle	Austria/ Switzerland	Buchle, a torch	Bachtold-Staubli (1927)
Aleya	Bangladesh		
Swetylko	Czechoslovakia	A lamp	Bachtold-Staubli (1927)
Lygtemand	Denmark	Lantern man	Bachtold-Staubli (1927)
Mosekone	Denmark	Woman brewing the marsh mists	J. Christensen (pers. comm.)
Liekkiö	Finland	Wandering unbaptised child. The 'flaming one'	Leach (1972)
Feux follets/ follets	France	Wispy lights	
Dickepoten	Lower Saxony, Germany	More than 12 variants known	Bachtold-Staubli (1927)
Irrlicht	Germany	general	Early name
Liam no lasoige	Eire	William with the little flame	Leach (1972)
Teine sionnic	Eire	Fox fire	Leach (1972)
Chir Batti	India	Gujarat State. Ghost light	
Cularsi	Italy	Bologna region	Beccari account
Hitodama	Japan	Human soul	Palmer's website
Dravlicht	Luxembourg	Wandering light	Bachtold-Staubli (1927)
Dwaallicht	Netherlands	Wandering light	Bachtold-Staubli (1927)
Ruskaly	Russia		Palmer's website
Irrbloss	Sweden	Fen-fire	Bachtold-Staubli (1927)
Eskudait	N. America (native American)		Leach (1972)
Witch fire	South Africa (Thonga tribe)		Leach (1972)

9.5.1   Bessel's Lights

This article was published in *Poggendorff's Annalen* for 1838 by the scientist F. W. Bessel and is based on a sighting from a boat over an adjacent peat bog near Wörpendorf. The bog had small water-filled depressions from which the lights arose...

These appearances were observed by me on December 2nd, 1807, early in the morning, on a very dark and calm night during which from time to time, a gentle rain fell. They consisted of numerous little flames which appeared over ground which was covered in many places with standing water and which after they had glowed for a time, disappeared. The colour of these flames was somewhat bluish, similar to the flame of the impure hydrogen which is prepared by the action of dilute sulphuric acid on iron. Their luminosity must have been insignificant, since I could not observe that the ground under one of them was illuminated nor that the great numbers of them which frequently appeared at the same time produced a noticeable brightness. A closer estimate of their brightness I cannot make, since the darkness of the night made my estimates of the distances of the flames very uncertain. Some of them, which seemed brighter than others, were estimated to be not more than fifteen or twenty steps distant, but this estimate is necessarily insecure.

As regards the number of flames which were visible at one time and as regards the period of their burning I cannot speak with certainty, since both conditions were quite variable. I can only estimate as some hundreds in number, and a quarter of a minute as the average period of their luminosity.

The flames frequently remained quite in one position, and at other times they moved about horizontally. When motion occurred, numerous groups of the flames seemed to move together. I remember that one of the groups of flames suggested the moving of flocks of water bird.

9.5.2 The Fulda Valley

The Fulda Valley is in Germany about 120 km NE of Frankfurt. The author of the original article is W. Loof and the article is published in Poggendorf's Annalen vol. 108 (1859) p. 656. Part of the article appears in ref. 13 translated into English...

The valley of the Fulda was covered by a heavy white fog, and a strong mouldy smelling vapour filled the air. Suddenly I saw a little flame scarcely two steps from me at the side of the road. I though I must be deceived but the moon was shining brightly and I was broad awake. To satisfy myself, I started toward the light, but when scarcely a foot distant it disappeared. But not a second had passed until I saw another, then a second, three, four others. All the little flames remained quite in one place and neither leaped nor danced. I observed that if the lights were not to disappear I must approach them very quietly, taking care not to set the air about them in motion. When I was very careful, I was often so fortunate as to bend over the little flames and observe their colour and form at a distance of not more than a foot and a half. They were little flames the size of a hen's egg, and they stood very quietly between the blades of grass. They were mostly of a greenish white light, and were fairly bright. I was able to seize some of them in my hand, but no heat was to be detected. If I waved a finger near them they disappeared at once. Many of them disappeared with a faint report, such as is made by the ignition of a bubble of phosphuretted hydrogen. Still, I must say that the air remained perfectly quiet.

A single flame seldom lasted longer than a minute and a half. The moon shone so brightly that I was able to read the dial of my watch. I could not have been deceived, for I observed the phenomenon very carefully and accurately. My eyes were completely clear, for I observed other objects about me and saw no lights between me and them.

9.5.3   Major Blesson's Experiment

In the Gorbitz Forest, Germany, the Major made the following observations in the 1830's; *During the day, bubbles were seen rising from [ferruginous water], and in the night blue flames were observed shooting from and playing over its surface. One day in the twilight, I went to the place and awaited the approach of night. The flames gradually became visible, showing that they burnt also during the day. I approached nearer, and they retired. Convinced that they would return again to the place of their origin when the agitation of the air ceased, I remained stationary and motionless, and observed them gradually approach. As I could easily reach them it occurred to me to attempt to light paper from them. For some time I did not succeed in this experiment, which I found was owing to my breathing. I therefore held my face from the flame, and also interposed a piece of cloth as a screen. On doing this I was able to singe paper, which became brown-coloured and covered with viscous moisture. I next used a narrow slip of paper, and enjoyed the pleasure of seeing it take fire. . .*

9.5.4 John Warltire's Letter

On 12th December 1776 Mr. Warltire, a lecturer in natural philosophy wrote to the chemist Joseph Priestley and this was copied into the appendix of Priestley's *Experiments and Observations of Different Kinds of Air*.[5]

> *Being about five miles from Birmingham, on the road to Bromsgrove, before day-light, the morning foggy in drifts, I observed with surprise, a great many Jack-a-lanterns, moving swiftly in an adjoining field, in several directions, from some of which there suddenly sprung up bright branches of light, something resembling the explosion of a rocket that contained many brilliant stars, if the discharge was upward instead of the usual direction, and the hedge and trees on the side of the road were illuminated. This appearance continued but a few seconds, and then the Jack-a-lanterns played as before.*

9.5.5 An English Graveyard

This abridged account was sent to the author by Kevin Spencer in 1998 and is reproduced with his permission: *As you can imagine, I remember it all very vividly. I saw the lights in June-July 1985. It was about 8.45 pm so it was just approaching dusk but still quite light and warm. The cemetery is on Middleton Road, south Leeds. There are two cemeteries on that side of the road and I saw the lights in the newer of the two, so the graves dated from about 1970 to the present day. At the time I was out for a walk and saw the lights from on top of a hill, looking down on the cemetery about 40 m away. At first I noticed a solitary orange light in front of the headstones. I stood watching the light but then it went out and another appeared at a different grave. I watched this for about 20 minutes as lights appeared then disappeared all across the graveyard in a seemingly random manner and no more than 3 or 4 in the whole graveyard at any time. The lights were orange in colour, very similar to streetlights. At the time I was only nine so I was quite scared and didn't go down to investigate further.*

On the same theme, R. Legg recalled that the Parish Church of Hampreston in Dorset is built upon a small mound of gravel only a metre or so above the meadows of the River Stour.[6] A Miss Billington wrote in 1883 that this church was regarded with supernatural dread because on certain nights of the year it was illuminated from the inside. The windows were lit with a brilliant white light and it was suggested at the time that this was due to the presence of *corpse candles*.

A few further descriptions warrant inclusion: In *Nature's Secrets*,[7] Willsford reported that '*... in moist places... pallid fires... forerunners of sultry heat in summer and wet in winter: they are usually observed to appear in open (i.e. mild) weather...*'. George Gough's edition of William Camden's *Britannia*[8] drew attention to the low grounds around Syleham, Suffolk where... *just by Wingfield, are the ignes fatui, commonly called Sylham Lamps, to the terror and destruction of travellers and even of the inhabitants, who are frequently misled by them*'... One wonders how today's villagers would respond to a modern report of this nature.

More recently, Robert Clark provided an interesting account of some sightings, most of which came from Germany.[9] They include a description given by Professor Knorr who saw 8 inch (20 cm) flames with a yellow centre and violet sheath apparently without heat. C. W. Schultze saw lights resembling burning matches that lasted for about 15 seconds spread over a distance of 2–4 boat-lengths in a flooded area. Herr Wisnewski saw near Hanover an area of rotting vegetation with flames resembling burning petroleum with the air smelling strongly of carnations. There was an account by Dressler from a boggy wood in Silesia where organic matter was seen to float to the surface buoyed up by gases. On a hot afternoon Dressler saw a floating 'bladder' explode and catch fire with a yellow-blue flame. The area was used as a burial place for cats and dogs.

The limnologist Leonard Beadle reported frightening sheets of flame from the ignited gas (methane) stirred up by the shallow-draft steam boat which at one time plied between Nyamsagali and Masindi Park on Lake Kioga, Uganda. In the same region, lights have been recorded by local villagers from marshland bordering Lake Nabugabo, near Lake Victoria in the late 20th century but were rarely seen.

9.6 *IGNIS FATUUS* IN THE UK

UK sightings of the *ignis fatuus* are shown in Figure 9.2. They have been placed into two categories. Those for reports

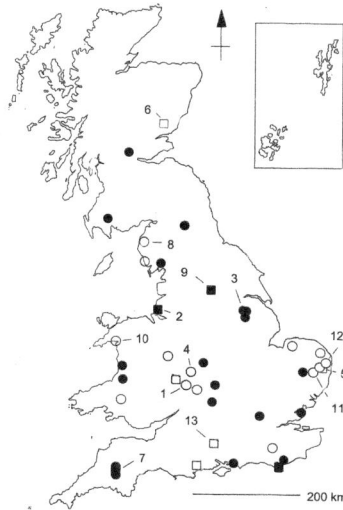

Figure 9.2   Distribution of the *ignis fatuus* reports in Britain. Open circles; pre-1900 records from sites not indicated as burials. Filled circles; post-1900 records. Open squares; pre-1900 records from burials. Filled squares; post-1900 burial records. Sites mentioned in text: (1) Powick, Hereford & Worcs.; (2) Blundellsands, Lancashire; (3) Blyton Carrs, Lincolnshire; (4) Bromsgrove, Hereford & Worcs.; (5) Bungay & Burgh St Peter, Norfolk; (6) Forfar, Tayside; (7) Foxtor Mire, Devon; (8) Lamplugh, Cumbria; (9) Leeds, West Yorkshire; (10) Morfa Bychan, Gwynedd; (11) Syleham, Suffolk; (12) Thurlton, Suffolk; (13) Wherwell, Dorset.

unconnected with graveyards or bodies and those known to be connected with them. Both categories have been subdivided into pre-1900 and post-1900 groups. A few clusters can be seen in both categories. There is a pre-1900 cluster in East Anglia, much of which lies just above mean sea level. Here there have been many reports in the past but few provide useful details. Another loose cluster occurs in the lowlands of Worcester including Powick, described by Allies in 1846. Among the single sites one of the most intriguing is that described by M. Lister in the *Philosophical Transactions.*[10] This refers to Morfa Bychan in Wales where *in the winter of 1694, the inhabitants were much alarmed by a fiery exhalation which arose from a sandy, marshy tract ... which injured much of the country by poisoning the grass, killing the cattle and firing hay and corn ricks.* The location of this site is interesting as it is close to the Mochras Fault described in Chapter 6.

Much of this earlier literature fails to give locations of the *ignis fatuus* and some cannot be shown in the figure. One of these relates to Cornwall and is unusual on account of its apparently practical application. It appears in the *Mineralogia Cornubiense* by William Pryce.[11] In Book 3 p. 112 in discussing ways of searching for minerals he writes: *Another way of finding veins, which we have heard from those whose veracity we are unwilling to question, is by igneous appearances or fiery coruscations. The Tinners generally compare these effluvia to blazing stars, or other whimsical likenesses, as their fears or hopes suggest, and search, with uncommon eagerness, the ground which these jack o' lanthorns have appeared over and pointed at. We have heard but little of these phenomena for many years...*

In Figure 9.2, two small groups can be seen post 1900. One is another lowland group in Lincolnshire that includes Blyton Carrs. These sightings are from the 1930s and are described briefly by Ethel Rudkin.[12] Finally there is a small upland group on Dartmoor.

Also plotted on the map are six sites relating to graves or bodies. These include Blundellsands in Lancashire where lights were associated with a buried body on the coast in 1902, and Eastbourne where a similar occurrence was reported in 1920. Another is Foxtor Mire on Dartmoor where, apparently, the body of an escaped prisoner was discovered. There are in addition the churchyards in Leeds in Yorkshire, Wherwell in Hampshire and Burgh St Peter in Norfolk where there are reports of 'corpse candles'. There is also an account of such lights near Forfar, Scotland in 1830. Such cemetery lights have been noted from several parts of the world including Japan and the United States.

Overall, sightings are seen to be widely distributed and do not appear to be influenced by the pattern of rainfall or geology. Sites range from ditches and small ponds to extensive areas of marshland although there is a dearth of information from Scotland. This may be a reflection of its low population density. Analysis of the data is difficult owing to the imprecise location of many of the reports but most are from fairly low altitudes, averaging 90 m in the UK. Some of the information on recent sites is derived from a questionnaire that the author posted to 114 likely looking locations close to wetlands throughout Britain. Among the few returns-less than 10% responded – only two sightings were reported. In addition to these records, there are several place-names that almost certainly refer to the *ignis fatuus*. A selection is provided in Table 9.2.

By analysing those accounts that provide what appears to be reliable information, a few inferences can be drawn regarding the nature of these lights. The phenomenon has been reported in all seasons, most often during winter and least often during spring (Figure 9.3a). This may not reflect the actual situation and may be related to the long winter nights increasing the chance of observation. Vegetation is also less prolific in winter, permitting better visibility of the ground. The majority of observations were made on calm days, and both fog and stormy conditions accompanied by thunder were mentioned several times (Figure 9.3b). The above discussion however points to the period June–September as the most auspicious time for sightings in Britain.

Most descriptions report blue flames but yellow flames, often accompanied by an outer blue sheath have also been noted. Reddish flames and white flames are also occasionally noted (Figure 9.4a). One of the most frequently observed phenomenon is flame movement. There are frequent reports of the flames skipping sideways with the individual flames separating and then coalescing. Stationary flames appear to be uncommon which is surprising considering the majority of observations were made on calm days (Figure 9.4b).

Table 9.2   UK place names with probable connections to marsh lights. OS = Ordnance Survey map.

Name	County	Nat. grid reference	Comments/source
Candle Dyke	Norfolk	63/4320	Broadland
Inglewhite Village	Lancashire	34/5440	Believed to be a direct reference to a marsh light appearing on the green
Lamps Moss	Cumbria	35/8104	OS
Lantern Moss Tarn	Cumbria	35/004056	OS
Lantern Man's Well	Suffolk	—	Dutt (1906)
Wisp Hill	Scotland	35/3899	OS
Wisp Wood	East Sussex	51/550343	1860 OS
Wisp Wood	Lancashire	34/5561	Forest of Bowland. Bibby (2005)

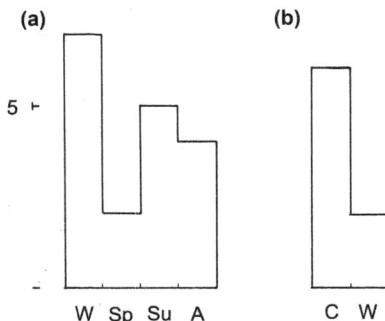

Figure 9.3   Frequency histograms of some *ignis fatuus* meteorological characteristics. (a) Seasons in which the phenomenon has been
observed. (b) Calm (C) or windy (W) conditions.

Although few observers were inclined – or able to determine
flame temperature it is clear that about half of the flames
examined were without any sense of heat (Figure 9.4c). The hot
flames tended to be blue and weak but data are too limited to
draw useful conclusions. Their appearance a short distance
above the ground is also evident and is also apparent in some of
the artwork described above. Flame height is seen to vary, but
the majority are in the 5–50 cm category. (Figure 9.4d) There is
little information on flame duration. While the displays often
last for several hours or even days, individual flames seem to
have short lifetimes of the order of a few seconds to 1 or
2 minutes. In three cases, small explosions occurred.

Odour or lack of it was infrequently reported. Two cases occurred
where the hot flames had a sulphurous smell. On one occasion a
smell resembling phosphane was observed and on another a smell
resembling the scent of carnations was noted (Figure 9.4e). Both of
the common sulphur-containing gases, hydrogen sulphide and
sulphur dioxide have distinct and strong odours and either one or
both may have been present. There are in addition many odoriferous organic compounds containing sulphur. There is little
useful information on the substratum but in most cases it appears
to have been wet soil, organic mud or peat, often covered with
shallow water. Only one account is definitely from marine waters
and this was of a buried body. The presence of water appears to be
important and the generally short duration of flames may result
from rising bubbles of gas bursting at the water surface.

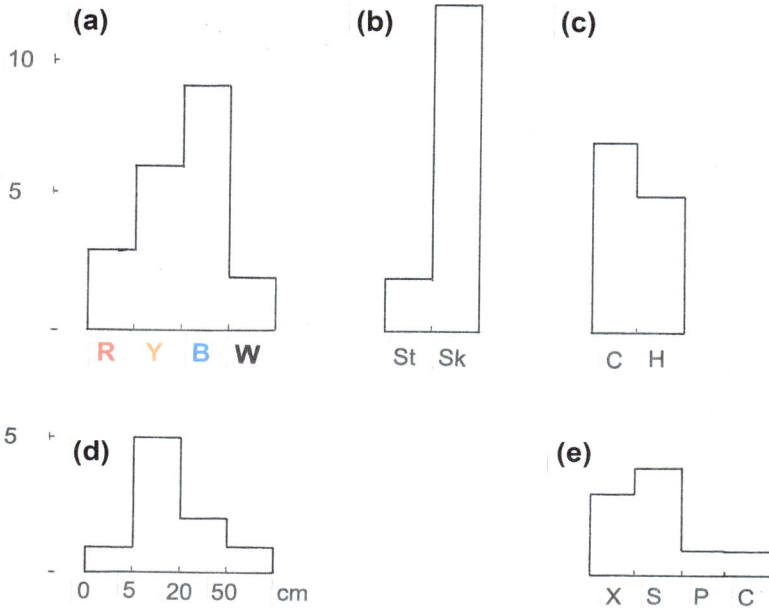

Figure 9.4   Histograms of *ignis fatuus* physical characteristics. (a) Predomin-
ant flame colour. (b) Stability. St static; Sk skipping motion. (c)
Flame temperature, cold or hot. (d) Flame height. (e) Odour: C
carnation; P phosphane; S sulphurous; X none.

It would appear that two types of phenomena have been witnessed, one with hot flames and one with cold. Blue or violet
flames are often the result of chemiluminescence while yellow
and red flames could be the result of incandescence. As in the
case of the candle (Chapter 1) both flame colours may occur
together as was noted in some of the above accounts. Incandescence signifies considerable heat while some chemiluminescent flames may be devoid of heat altogether. In the latter,
convection of air around the flame should be minimal giving a
more rounded or irregular shape to the flame but the shape is
not described in sufficient detail to make a distinction. The apparent movement of the flames presents an enigma as it implies
that their fuel must also travel over the ground. Perhaps the effect is illusory and a form of autokinesis. Overall the data are too
sparse to permit the recognition of more than one fundamental

type of phenomenon. However as the next section will show, there is no shortage of possible explanations.

9.7 EXPLANATIONS

Allan Mills grouped *ignis fatuus* observations into three classes of explainable phenomena: combustion, chemiluminescence and bioluminescence.[13] In the last category he drew attention to the writings of William Derham, who many years ago dismissed bioluminescence as an explanation. Derham, in 1730 had described an *ignis fatuus* as *frisking around a dead thistle in a field* that he was able to witness at close quarters.[14] Mills could find no connection with bioluminescence, such as that found in glow-worms or fungi and gave more weight to chemiluminescence citing a number of previous studies. He concluded that multiple flame occurrences were the norm and their frequent extinction and rekindling probably explained folklore descriptions of their dancing from place to place. In the following sections, attention is first given to combustion followed by chemiluminescence and bioluminescence.

9.7.1 Combustion of Marsh Gas

Marsh gas, or methane, is a common flammable product of organic decay and has been previously noted. In marshland it is released to the air in rising bubbles or by diffusion into the air. It is the most important constituent of marsh gas (Table 9.3).

The auto-ignition temperature of pure methane in air is 580 °C. In marsh gas it may be higher since here it is mixed with nitrogen and carbon dioxide, gases that are not combustible.

The methane found in wetlands is almost entirely a product of organic matter breakdown resulting from methanogenesis. This occurs in two stages: first organic matter is metabolised to fatty acids *via* hydrolysis and fermentation. Next, the fatty acids are reduced to methane by methanogenic bacteria, either by the reduction of carbon dioxide by hydrogen or by fermentation. The latter is more general and also yields carbon dioxide. High levels of other common electron acceptors (mainly oxidisers) such as nitrate, iron(III) and sulphate can inhibit methane formation, depending on their concentration. Methanogenesis can only occur in the absence of free oxygen and it tends to be higher in

Table 9.3 Typical composition of marsh gases.

Component in molar%	Freshwater marsh	Brackish marsh
Methane	68	73
Propane	<0.01	0.02
Hydrogen	0.05	0.08
Carbon monoxide	0.5	0.5
Carbon dioxide	8.2	15
Nitrogen	22	10
Hydrogen sulphide	Trace	0.4

wetlands. This is because oxygen is sparingly soluble in water. Where the water table is low so that the soil becomes aerated, much methane is subsequently oxidised by bacteria known as methanotrophs. As much as 90% of the methane may be lost in sediments and soils in this way. Photosynthesis in shallow waters may also result in faster methane oxidation rates during the day than at night owing to oxygen production. Plants can take up methane *via* their roots by diffusion and release it through their leaves and stems *via* their internal ventilation system. More than 90% of the methane produced may escape by this route in wetlands rich in plants.

Despite the activities of methanotrophs, in lakes, methane loss rates to the atmosphere (often termed evasion rates) are higher in shallow than in deep water. This is due to the slow rate of diffusion of the gas through water. Typical values for shallow British waters are in the range of 150–300 cc methane $m^{-2} day^{-1}$, with more at higher temperatures.[15] The rate of gas escape therefore depends on a range of factors. Temperature affects microbial activity and to a small extent the diffusion rate of the gas. The organic matter content of the soil is important as it varies from practically zero to almost 100%. The soil pH affects the metabolism of these microbes and the optimum is around 6. The amount of water that is present affects both the rate of methane diffusion and the activity of the methanotrophs. Studies of a US peatland showed that rainfall increased the hydrostatic pressure within the peat and this affected the evolution of gas. Changes in air pressure could also be significant.

To summarise, a warm, wet, but not water-saturated soil of pH 6, rich in organic matter should provide ideal conditions for its formation. If in any quantity, methane burns with a luminous yellow flame. Its spontaneous combustion provides a considerable challenge owing to its high auto-ignition temperature.

Hydrogen gas is also formed in organic sediments but most is rapidly recycled by microbes, so it is a minor component of marsh gas. Carbon monoxide is also formed and lost in a similar way. Both of these gases are flammable. Carbon dioxide is an important component of marsh gas and evasion rates from bogs are around 500–1000 cc per square metre per day.

9.7.2   The Diphosphane Hypothesis

The involvement of phosphorus in marsh lights was originally proposed by Alessandro Volta. However, Robert Boyle had previously drawn similarities between it and the Will o' the wisp around 1680. Boyle became obsessed with phosphorus for several years after seeing a demonstration in London. Perhaps the earliest recorded preparation of diphosphane was by the chemist Phillipe Gingembre who in 1783 boiled white phosphorus with a solution of caustic soda and obtained a gas that was spontaneously flammable in air. A decade later a similar gas was found by adding water to calcium phosphide. The actual cause of combustion was unclear because an almost identical gas had previously been obtained by heating phosphorous acid that was *not* spontaneously flammable. It was left to the French chemist Louis Thénard to discover that the flammable gas was actually a mixture of at least two phosphorus hydrides, phosphane (PH_3) which is reasonably stable plus traces of the spontaneously flammable hydrogen disphosphide (P_2H_4), now called diphosphane.

Today, the routine laboratory preparation of diphosphane still employs calcium phosphide but the product is impure and the apparatus complex owing to the flammability of the product. It is an unstable compound that is decomposed by light and is difficult to keep (Figure 9.5). In times past cargoes of ferrosilicon have spontaneously exploded and this is thought to have been caused by emissions of hydrogen phosphides present as an impurity.

Diphosphane is a liquid at room temperature but its boiling point is low so it readily volatilises with decomposition. Its spontaneous combustion is dramatic with the emission of a

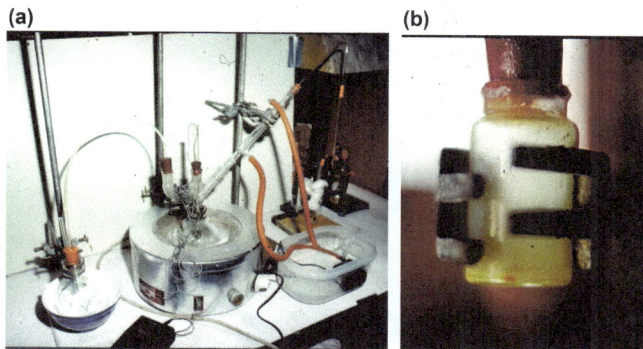

Figure 9.5 (a) The traditional method of preparing diphosphane. The flask on the left is a carbon dioxide generator used to protect the phosphorus hydrides from oxidation. The centre flask placed in the heating element contains lumps of calcium phosphide onto which a dilute mineral acid is poured to release the hydrides. Temperature is maintained at 40 °C so that the diphosphane will vapourise, passing through the straight condenser to remove water vapour. The hydride condenses in a small tube placed in a freezing mixture. (b) The pale yellow liquid hydride collected in a small bottle accompanied by white fumes of phosphorus pentoxide.

bright yellow-white light accompanied by noxious fumes of phosphorus pentoxide. Conflagration is most readily observed by dropping some calcium phosphide into water (Figure 9.6). The reaction is accompanied by an unpleasant smell resembling rotting fish caused by the unreacted hydrogen phosphides. Concentrations down to 1 part in 500 diphosphane are capable of igniting other flammable gases at room temperature. The compound is structurally similar to hydrazine, N_2H_4, another unstable compound.

Few spontaneously igniting gases were known in the 19th century and it is natural that a connection should be made between this gas and the *ignis fatuus*, although as we have seen, actual combustion of gases is not always observed. There are no *ignis fatuus* reports combining the observation of white fumes with a fishy smell although in one case, 'smoke rings' have been reported. Nevertheless, if the gas were present in traces these characters could be masked. Mills injected impure phosphane into a current of natural gas and found that a bright green luminescent flame could be produced without ignition. This gas would probably have been contaminated by diphosphane but he concluded that the flame colour was wrong and there was still

Figure 9.6 Spontaneous combustion of diphosphane present as an impurity of phosphane. The hydrides were produced by dropping calcium phosphide into water in the writer's garden. Emission is continuous in the visible region of the spectrum and is accompanied by some UV radiation.

the characteristic fishy smell which almost all marshlight reports fail to mention. Nevertheless, given a low enough concentration, diphosphane must remain a possible explanation.

Milne-Edwards concluded in 1863 that the luminescence of decaying fish resulted from diphosphane formed by the decomposition of organic phosphorus compounds but there were no supporting experiments.[16] The faint illumination of sandy beaches at night was also thought to result from this. A more likely explanation is the growth of luminescent bacteria although there have been reports of oily fish meal spontaneously combusting.

Large numbers of bacteria and fungi live in the soil. Recent studies analysing their DNA indicate that a great variety of types occur forming complex microbial communities. Organic soils tend to support the largest numbers and wet soils tend to support more anaerobic forms. All bacteria contain a significant amount of phosphorus and are, along with other microbes such as fungi, one of the main sources of phosphorus in soils. So, is there is a connection between the phosphorus hydrides and bacteria? Recent studies using bacterial cultures show that there are.

The Russian biologist K. Rudakov claimed to have isolated soil bacteria that could carry out a reduction of phosphate to phosphane. This was later supported by redox data and the detection of reduced phosphorus compounds in both sewage treatment works and the bacteria *Clostridium* and *Escherischia coli*.[17] The phosphane was trapped in bottles containing nitric acid which oxidised the gas to phosphate so that it could be detected by conventional analysis. In 1995 Devai & Delaune reported it from subtropical marshy soils and estimated evolution rates of up to 6.5 nanograms per square metre per hour.[18] It has also been found occasionally in volcanic gases and human flatus.

Phosphorus hydrides are difficult to quantify without special equipment. The biologists J. Burford & J. Bremner had previously reported the occurrence of phosphane in soils using a sensitive gas chromatographic method, but they could not confirm previous reports of phosphane formation in waterlogged soils.[19] They did note that phosphane is sorbed by soil components and may thus remain undetected. They reported on a new chemical trapping method employing a solution of alkaline potassium permanganate which converted phosphorus hydrides into phosphate and estimated that as little as 20 nanograms of phosphorus per gram of soil can be detected in this way. Then in 1993, two German chemists, Günter Gassmann and Dieter Glindemann, found convincing evidence for the existence of phosphane in the natural environment using gas chromatography.[20] These chemists managed to find traces of phosphane in the guts of slaughtered animals, a highly anaerobic environment where oxygen concentrations would approach zero. They also found it in human excrement and it appeared that the levels increased as the amount of phosphorus in the diet increased. Levels of up to 200 nanograms per kilogram were found together with traces of diphosphane. These are extremely small amounts and are a hundred times lower than those achievable using the gas-trapping method noted above. They also found a positive association between phosphane and methane production.

More recently Jenkins and co-workers detected phosphane in several cultures of anaerobic bacteria.[21] These were incubated in a range of nutrient media and up to 720 picograms (10^{-12} g) phosphane per ml of gas was found after an incubation period of 8–16 weeks at 34 °C. No other phosphorus hydrides were detected but methane, hydrogen sulphide and some volatile sulphur compounds were found. Animal faeces have also been found to be a good source of phosphane. The bacteria responsible were *Escherischia coli*, and several species of *Clostridium* and *Salmonella*. The microbiologists noted that several strains contained an enzyme known as formate hydrogenase. This enzyme is able to catalyse a strong reduction with a redox potential of −0.569 volts. Another enzyme present in some *Clostridium* species and known as carbon monoxide dehydrogenase can carry out reduction at an even lower potential of −0.622 volts. A strongly reducing environment would be needed to remove the oxygen from phosphate. Bacterial sulphate reduction needs such an environment, but a more extreme situation would be required to reduce phosphate. This is because the phosphate atom is small when compared with the sulphate atom and the oxygen atoms are more strongly bound to it.

There is then, little doubt that phosphorus hydrides are present in soils and are of microbial origin. These hydrides would not be formed directly from phosphate $(PO_4)^{3-}$ but from a more reduced ion such as phosphite $(PO_3)^{3-}$ or hypophosphite $(PO_2)^{2-}$. Both are known to occur in organisms. Although phosphane has been used as a fumigant for animal feed it is unlikely to persist in this situation and was not available before the 20th century. It can therefore be discounted as a phosphorus source for the *ignis fatuus*.

There have been several attempts to produce a Will o' the wisp artificially. Alan Mills set up his own experiment by placing some peat, soil and compost in a flask with additions of bone meal, ammonium phosphate dried milk and fish.[22] Although the resulting gases were highly odoriferous he failed to observe an *ignis fatuus*. The author also set up some flasks containing 80 grams of a wet sand and garden soil mixture plus 7.5 grams of minced beef. The flasks were injected with a small amount of phosphate containing the radioactive isotope ^{32}P. They were incubated at room temperature in the dark for a month. Radiotracing is a sensitive method for the detection of phosphorus hydrides but none was detected in the copper sulphate trap. It may be that the phosphate-reducing bacteria were slow starters and had not attained sufficient numbers to allow detection.

The research so far has shown that phosphorus hydrides are produced preferentially under moist anaerobic conditions. Dry soil contains oxygen although small anaerobic pockets occur. In these and in water-saturated soils when oxygen levels are low phosphane and diphosphane could be formed and survive long enough to diffuse into the air. Oxygen is sparingly soluble in water and is soon consumed by respiring microbes. Phosphorus hydrides are poisonous to most forms of life so why would bacteria produce them? Perhaps some species use it as a form of defence or to reduce competition. It is also possible that it is the result of a side-reaction originating from the reduction of other compounds such as sulphate. Reactions of this type are well known. Oxygen scavenging from the phosphate for use by microorganisms is also a possibility since phosphorus levels in decaying organisms, especially bacteria and vertebrates are quite high.

Phosphane has an intensely disgusting smell which is detectable in small concentration. It is surprising that if the phosphorus hydrides are involved in the *ignis fatuus*, this would be mentioned more often in reports. If diphosphane has an odour, it has not been reported, but this is not surprising given its instability and frequent contamination with phosphane. The two compounds, being closely related, would be expected to occur together on most occasions. This observation is not new, and Marcel Minnaert mentioned that a mixture of phosphane and hydrogen sulphide, when ignited, is colourless and odourless.[23] There has been little research on the topic and the proportions of the two gases are probably important. Hydrogen sulphide, also a strongly smelling gas, is sometimes detectable in marsh gas.

Elemental phosphorus can also vaporise in warm air and is well known for producing spectacular cold flames in the laboratory (Chapter 1). It is just conceivable that this could also be produced through the microbial oxidation of the phosphorus hydrides. After all, some of the world's sulphur deposits were formed by the bacterial oxidation of hydrogen sulphide. Judging by the extremely low concentrations of the bacterially-produced hydrides, however, it is unlikely to be capable of being produced in sufficient quantity to create luminescence. Although diphosphane has received much attention with reference to the *ignis fatuus*, it has also been found that phosphane itself can spontaneously oxidise in the laboratory. Experiments mixing low concentrations of the gas with nitrogen containing a trace of oxygen give evidence of oxidation through chemical analyses of the products. These have shown that phosphane can be oxidised at room temperature in a mixture containing just 1% of oxygen. A chain-building and chain-breaking process has been suggested, similar to that operating in the oxidation of silane. The process sheds some light on the likely oxidation process in the phosphorus hydrides. Finally, preparations of a substituted phosphane have been found to self-ignite. The material in question is a gas, methyl phosphane, CH_3PH_2. It is the simplest of the many organo-phosphorus compounds but has not so far been found outside the laboratory. There also exists a bewildering array of organo-phosphorus and metal–phosphorus complexes such as the phosphanides. Some of them may yet be shown to play a role in spontaneously combusting natural emissions.

Having investigated the phosphorus hydrides in some detail, it is useful to pay some attention to the concentration of phosphorus in soil. Being a complex mixture of organic and inorganic matter, the soil presents problems for chemists who wish to know the distribution of elements such as phosphorus within it. Numerous schemes have been devised to separate the soil components so that this may be achieved. A perfect separation has proved impossible leading to the recognition of soil 'fractions' instead. These fractions attempt to separate the inorganic and organic components. Despite the uncertainties, much progress has been made and it can be stated without question that by far the greatest fraction of the soil phosphorus exists as phosphate (PO_4). Of particular relevance is that proportion of the soil phosphorus that is available to microorganisms. This fraction unfortunately cannot be determined with precision owing to the great variety of microbes present. Some microbes can obtain their phosphorus directly from mineral grains such as apatite, a form of calcium phosphate. Others secrete enzymes capable of removing phosphorus from organic compounds within the soil. It is thus necessary to fall back upon a more fundamental measure, the total phosphorus content. For British agricultural soils this averages about 700 parts per million by weight. It is considerably higher than the global average, due partly to a long history of fertilisation and improvement in this country. The phosphorus content of organisms is also of interest and some relevant figures are found in Table 9.4.

Bacteria are seen to have a high phosphorus content. This is because some species store phosphorus in excess of their immediate need. It is also high because bacteria have thin cell walls. Cell walls of most plants contain cellulose. This does not contain phosphorus so the overall phosphorus content of plants is effectively diluted out. Overall however, most of the soil phosphorus is present in the non-living material. A square metre of reasonably rich soil will contain about 250 grams of live bacteria. These will contain about two and a half grams of phosphorus. With a fertile soil depth of 30 cm and a phosphorus content of 700 ppm, the total phosphorus content is around 300 grams per square metre, so the bacterial population will only contain about 1% of the phosphorus present.

Since buried bodies have been shown to be associated with the *ignis fatuus* sightings their phosphorus content is also of interest. Terrestrial vertebrates have a high phosphorus content owing to their well-developed skeleton, much of which consists of calcium phosphate Table 9.4.

The microbial activity of soils shows a strong dependence upon weather and climate. Permanently wet soils require a regular source of water if the temperature is far from freezing. This is not necessarily the case if the water is cold but then, bacterial activity will be low. In general rainfall needs to be reasonably high at least for part of the year. This eliminates large areas of the planet although these regions will also tend to be less inhabited.

If marsh lights are a product of biological activity then they will be influenced by temperature. In the higher latitudes where the seasons are well marked, soil temperature varies in a fairly predictable way. For the northern hemisphere they are warmest in

Table 9.4   Representative values of the phosphorus
content of some organisms and soil.

Material	P content parts per million (dry weights)
Bacteria	10 000
Vegetation	1600
Fish	5000
Human	16 000
Soil	100–700

late summer and coldest in late winter. Dry soils respond rapidly to solar heating compared with wet soils since water has a high heat capacity. Furthermore, the effect of solar heating diminishes with soil depth and at depths over 4 m the effect of seasonal heating is greatly reduced. The presence of vegetation also influences the temperature. Vegetated soils are cooler during the day and warmer during the night compared with bare soils. Overall, it is clear that the relationship between soil temperature and season is complex although some generalisations can be made. In Britain, soil temperature reaches a maximum in July–August. There is a skew in the curve for the later months caused by heat retention. Thus autumn soils will be a little warmer than spring soils. The pattern for marshland soils should follow a similar trend. The higher heat capacity of the saturating water should enhance heat retention during autumn to a small degree. If there is a link between microbial activity and marsh lights then late summer and autumn should provide the most sightings.

9.7.3   Other Self-igniting Gases

If bacteria are capable of reducing phosphate to hydrogen phosphide then they may also be capable of reducing borates to hydrogen borides and silicates to hydrogen silicides, both of which include spontaneously combustible gases.[24] The mineral quartz composed of silicon dioxide is ubiquitous in the environment, but there is no evidence that bacterial reduction takes place. Most ground waters contain traces of silicate, the average being just a few parts per million. Some silicon is used by plants so there exist effective pathways for its uptake and use. Grasses for example contain considerable amounts of amorphous silica and are of course an important component of many composts. Bacteria no doubt can also utilise it if required. The situation with boron is different. Although borates are scarce in natural waters, many of the salts are soluble. In addition the simplest of the hydrides, diborane is rapidly destroyed by water so it could not survive in soil. Pentaborane however is less reactive and could exist in soils and then be exhaled into the atmosphere. In this context it might be relevant that borates have been detected in some graveyards but there is no obvious explanation as to why this is so. It has also been suggested that organometallic compounds may be responsible for the *ignis fatuus*. A good number of these are spontaneously inflammable but their occurrence in nature has hardly been investigated.

9.7.4 Hydrocarbon Chemiluminescence

Oxidation of the hydrocarbon pentane, C_5H_{12}, may result in chemiluminescence. It appears that the gas emits a blue glow near 220 °C when in low concentration in the presence of atmospheric oxygen. Such 'cool flames' have been attributed to the formation of excited molecules such as formaldehyde. This can be formed from the reaction of the methoxy radical with hydroxyl: $CH_3O + OH = H_2O + HCHO^*$ produced during a complex oxidation process. There is also some evidence that similar reactions occur in the marsh gas methane. It still requires a substantial rise in temperature though this might be achievable in sizeable piles of rotting vegetation, a location where the *ignis fatuus* has sometimes been observed. James Barry speculated that lightning strikes could release methane in sufficient amounts to support some form of combustion.[25] He cited work by P. Hubert who triggered lightning strokes and found luminous globes developed near ground strikes suggesting they may be due to the release of gases.[26]

Both plants and soil emit volatile substances that may be capable of combustion or chemiluminescence. The 'burning bush', *Dictamnus* has already been noted. Volatiles released by soils and similar surfaces after rainfall and known as petrichor include flammable organic compounds such as geosmin.

However they are not known to self-ignite or chemiluminesce in the air.

9.7.5   Luminescent Bacteria and Fungi

There is no doubt that some reports of the *ignis fatuus* refer to bioluminescent or fluorescent organisms. When Paul Hentzner returned home from Canterbury in 1598 he wrote that *there were a great many Jack o' Lanthorns so that we were quite seized with horror and amazement.* These were no doubt glow worms, once fairly common in this part of England. Other reports may refer to the common toadstool *Armillaria mellea.* Walking over humus and rotting wood at night often reveals its softly glowing mycelium. The luminosity of *Armillaria* has been known for centuries and it is one of the few British fungi to display it. In 1652, Olaus Magnus described how people in the northern Europe placed pieces of rotten oak bark, probably infected with this fungus, at intervals along paths to assist night-time travellers. This fungus attacks a wide range of trees and the actively growing mycelium may show a fairly strong luminosity.

A more intriguing source of luminosity in wetlands is bacteria. Luminous microbes are rarely reported from freshwaters. This may be connected with the high sodium requirement of many luminous species. Sodium levels in freshwaters are almost 2000 times lower than in sea water where luminous species are often abundant. Freshwater luminous bacteria are nevertheless present either free-living or within symbioses with other organisms. Most reports of free living forms come from the surface of dead and decaying animals or their remains.

Luminous bacteria have been seen on salmonids, fish that often switch from saline to freshwaters during breeding. On rare occasions freshwater shrimp have been found with luminous bacteria but there seem to have been no reports from northern Europe. Luminous bacteria are also known from infected midges and are likely to be the luminous agents in mayflies and ants although rarely reported. Some earthworms are luminous, *Diplocardia* and *Enchytraeus,* being well known examples but here bacteria do not appear to be involved. In his *Natural History of Staffordshire* of 1686, Robert Plot described his horse breaking through a dark spongy earth on Archer Moor revealing 'many embers'. They were probably caused by fungal mycelia but earthworms could also have been responsible.

On a more macabre note, luminosity has been reported on human corpses. A little known study on glowing cadavers was published by D. & R. Cooper working at the School of Anatomy and Medicine in Borough, London, in the nineteenth century.[27] Over time the glow of a corpse was seen to extend over the bones and tendons. Microscopic examination of the oily matter revealed the presence of minute globules darting from side to side. This was probably due to Brownian motion caused by their small size. The Coopers were unable to confirm that they were of living organisms although they were almost certainly luminous bacteria. They also noted that the phenomenon was known from the dead bodies of birds, dogs and cats in the warmer months of the year. Some of the *ignis fatuus* accounts from the Norfolk Broads may relate to these bacteria growing on dead fish floating in the lakes and ditches.

Norfolk Broadland, where there have been many reports of marsh lights,[28] is low-lying and only a few metres above sea level. At high tide, sea water being more dense than fresh water can flow inland as a separate, lower layer for several kilometres. If these contained luminous marine bacteria or protozoa such as *Noctiluca*, the waters would exhibit a bluish glow and could easily be mistaken for the *ignis fatuus*.

It is also conceivable that bioluminescent bacteria could enter a bubble of marsh gas and become expelled at a wetland surface giving rise to a marsh light. This would be a cold light with a bluish-green colour but the question arises as to how many bacteria would be needed to give visible light. For a bacterium emitting 1000 photons of light per second, a reasonable estimate, a well-sighted person standing 3 m away from a 5 litre 'bubble' of bacterial aerosol would see a faint image if around two million bacteria were present. This is a small number of bacteria but the production of aerosol is questionable. There is no obvious way in which this might be achieved.

Luminous owls have been reported in parts of West Africa and elsewhere. Their virtually noiseless and swift flight and nocturnal activities make them prime contenders for some of the accounts. In a fascinating article based upon first-hand observations in Britain, H. Robinson came up with three suggestions.[29] He thought that the feathers may become contaminated with wood fragments containing luminous fungi by the bird rubbing against rotting wood. Alternatively he suggested the occurrence of a luminous feather fungus or secretion brought on by disease. Nobody seems to have investigated these ideas further. Another possibility is fluorescence. Feathers of budgerigars and other birds are known to fluoresce, the excitation provided by the sun's rays. In the case of the budgie the fluorescence acts as a sexual attractant. White sheep's wool can also luminesce faintly in UV light and sheep often appear to brighten at dusk.

Edmund Harvey, the author of the influential *History of Luminescence*[30] admitted to have never encountered marsh lights and wondered whether they were no more than low-lying mists faintly illuminated by sunlight.

9.8   DEMISE OF THE WISPS

The many references to waterlogged ground suggest that the *ignis fatuus* frequently originate within the damp or water-saturated soils. Due to their importance in agriculture, soils have been well studied so that much is known of their properties. However most soils are not permanently saturated with water. To be agriculturally valued, soils generally need to be well aerated allowing oxygen to penetrate plant roots although there are some well-known exceptions such as rice paddies. Aerated soils require a degree of porosity to allow the gases to freely circulate within them and this is lacking in a waterlogged soil. Therefore, fields which have been flooded cease to have well aerated soils, and the free oxygen within them is soon removed by the soil microbes and plant roots. These soils lose much of their microbial activity as the oxygen level declines. From a farming perspective, the flooded soils of marshland result in low productivity and difficulty of working. This has led to the initiation of extensive drainage schemes to improve them.

Britain like much of Europe has a long history of land drainage. The Romans cut ditches on Romney Marsh in Kent and there are many references to drainage in Medieval times. In the Norfolk Broads, marsh drainage dates back to Tudor times if not earlier.

Efforts appear to have accelerated in the mid to late 18th century, aided by the construction of windmills to pump water away from the low-lying land, making it suitable for cattle grazing. At the end of the 18th century, tile-and-brick land drains began to replace the traditional 'faggot and stone method' of field drainage and their emplacement continued apace up to the 1850s. This rapid increase was due to the expanding population following the industrial revolution. Drainage continued well into the 20th century and it became largely mechanised by the 1940s. The laying of land drains represents a huge investment in labour and materials and must be one of the main reasons why the *ignis fatuus* is so rarely seen today. These losses of wetlands have been well documented. For example, Dr C. Sinker of the Field Studies Council observed that by 1952, the Cheshire and Shropshire mosses (wetlands) had lost 80% of their area to land improvement and peat digging[31] (Figure 9.7). A similar situation has occurred with farm ponds. Many were seen to represent unproductive land and were filled in to increase the arable acreage. In Lancashire, the botanist E. Greenwood found that in an area of 100 km² about 733 ponds representing 39% of the total were lost between 1951 and 2000.[32]

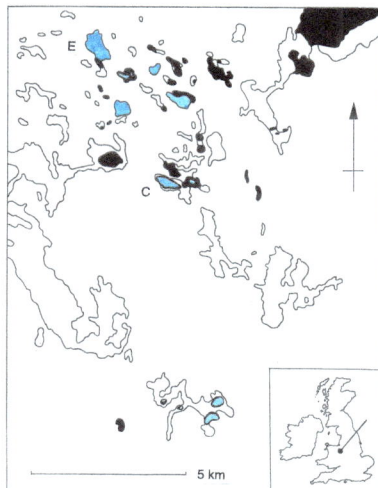

Figure 9.7 Loss of peatland in north Shropshire, UK, up to 1952. Existing peatland is shown in black, lost peatland unshaded. Waterbodies shown in blue. C Crosemere; E Ellesmere. Reproduced from ref. 31 with permission.

In other parts of Britain, particularly the south-east, ponds and marshes have also been lost due to water abstraction.

Stories as far back as the 19th century show that marsh lights were formerly more plentiful. In 1884 the East Anglian wherrymen has already noted that the lights were less common than they once were.[1] Night fishing for eels was frequently practised at that time yet there appear to have been few sightings. A number of brief articles appeared in the British journal *Weather* in the early 1950s. They commented upon the lack of recent reports and requested readers to forward any sightings. None was forthcoming suggesting that by this time, they were becoming rare.

More recently, several surveys have been undertaken to determine their frequency and distribution. Lars-Erik Astrom sent a questionnaire to several hundred Swedish local history societies enquiring about recent (1990s) sightings. He obtained few positive results and thought that modern lifestyles may be responsible for this lack of information. Alan Mills also undertook extensive enquiries without result.

In the author's survey, noted above, one farmer from Somerset had not seen it in 37 years and thought the land was too well drained. The questionnaire revealed just two further sightings, one at Blo Norton in Norfolk and another near Wigtown in Dumfries & Galloway. In Denmark, Jurgen Christensen informed me the phenomenon is no longer reported in the country. Further afield, Professor Zhang Zhao-hui of Guizhou University told me that occurrences of the phenomena in China are nowadays rarely reported compared with the past.

Few venture out into the dark nowadays. In the UK even fairly remote villages sometimes have night-long street illumination, and unlit lanes are often used as shortcuts by fast motor vehicles making them unsafe for pedestrians. Needless to say, it does seem that marsh lights really are less common than formerly. With the known association between the *ignis fatuus* and dead bodies the most likely places where the *ignes fatui* will be encountered nowadays are in graveyards. Admittedly, lurking in graveyards after dark is not everyone's cup of tea although they provide a safer environment than many wetlands.

Despite the considerable effort undertaken by researchers, marsh lights remain an enigma. As in the case of other unusual phenomena, their rarity and unpredictability represent a considerable obstacle to their investigation. The involvement of reduced phosphorus compounds appears most likely and is worth following up. A few suggestions as to how this might be achieved are provided in the next section.

9.9 SEARCHING AND RESEARCHING

The author has spent a good many night hours wandering through dark marshes, exploring graveyards and ditches without success. These visits were not without interest and some experience was gained as a result. Some simple equipment should accompany the explorer so that in the event of a sighting, an effective means of sampling can be undertaken. If the observer is fortunate enough to contact a source, some simple chemical analyses can be undertaken. Sampling the products of oxidation can be done using a 'cold finger' (Figure 9.8). This is a clean

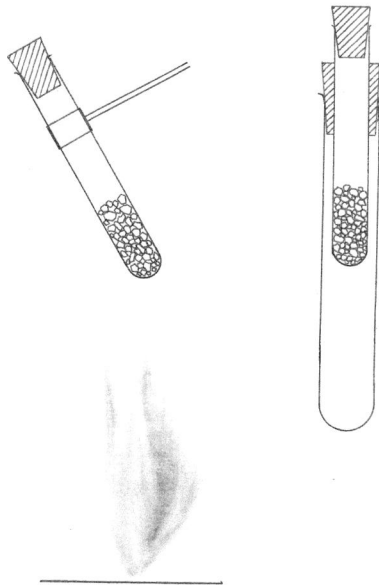

Figure 9.8 The 'cold finger' method of collecting condensable gases. See text for details.

Pyrex test tube measuring around 20 mm wide. It is one third filled with crushed ice. The ice is kept in a small thermos flask until required. The tube should be held over the *ignis fatuus* so that any oxidation products will condense on the surface. After a measured period of time the finger is fitted into a large test tube of about 30 mm diameter with a bung to prevent contamination. It must be kept vertical to prevent the melting ice from spilling out. Ideally, a second cold finger should be exposed for the same period of time at some distance from the site to act as a control. This could also be used to determine whether any water vapour had been produced by the lights. The condensate could contain water-soluble oxidation products such as phosphate, nitrate and sulphite ions. There are simple and sensitive methods for analysing these. Chemical test kits for water analysis are worth considering if a laboratory is not available. Temperature measurement is best done with a thermistor as it has a low heat capacity. For inaccessible sources, thermal imagers are available but expensive. A digital camera can be used to record the event.

Late summer to early autumn appears to be the best time of the year to search in northern latitudes. Unfortunately most wetlands in choked with vegetation at this time and the writer has often been confronted with metre-high reeds obscuring the view. Ditches are particularly difficult to examine unless there are crossing places. A dark-adapted eye is needed to search since some of the lights are reported to be faint. Wandering about on moonless nights with a dimmed torch is probably ideal and the researcher may be rewarded with some interesting experiences. While visiting the North Kent marshes on a dark winter's night, we were delighted to see flickering lights about a mile distant. As we drew closer we could see some regular movement until eventually, naked figures appeared dancing around a small bonfire among a ruined building. We decided not to quiz the participants about marsh lights and returned home amused but disappointed!

FURTHER READING

J. Bord and C. Bord, *Modern mysteries of Britain*, Harper Collins, New York & London, 1987.

C. M. Cade and D. Davis, *The taming of the thunderbolts*, Abelard-Schuman, London, 1969.

W. R. Corliss, *Handbook of unusual natural phenomena*, Anchor Books, New York, 1983.

J. Emsley, *The shocking history of phosphorus: a biography of the Devil's Element*, Cambridge University Press, 2000.

E. A. Fitzpatrick, *Soils, their formation, classification and distribution*, Longman, New York, 1980.

J. Garratt, Will-o'-the-wisps and hot compost, *Chem. Rev.*, 1994, 3, 30.

K. Haupt, *Der Irrwisch*. Sagenbuch der Lausitz. Vol. 1. Leipzig. Translated by D. L. Ashimat, 1862.

W. C. Hazlitt, *Dictionary of Faiths and Folklore*, Reeves & Turner, London, 1905.

S. B. Palmer, Will-o-the-Wisp – The Archives, retrieved 2024, https//Inamidst.com/lights/wisp/.

D. P. Penhallow, A blazing beach, *Science*, 1905, **22**, 794–796.

F. Sanford, Ignis Fatuus, Sci. Mon., 1919, **9**, 358–365.

REFERENCES

1. W. H. Barrett and R. P. Garrod, *East Anglian Folklore*, Routledge & Kegan Paul, London, 1976.
2. W. W. Sikes, *British Goblins: Welsh folklore, fairy mythology, legends and traditions*, Sampson Lowe, London, 1880.
3. G. C. Davies, *Norfolk Broads and Rivers*, Blackwood & Sons, Edinburgh & London, 1884.
4. J. Allies, *On the Ignis Fatuus or Will o' the Wisp and the Fairies*, W. Edwards, Worcester, 1846.
5. J. Priestley, *Experiments and Observations on Different Kinds of Air*, London, 1779, 3, 1, 368
6. R. Legg, *Mysterious Dorset*, Dorset Publishing Company, Sherborne, 1987.
7. T. Willsford, *Nature's Secrets*, N. Brook, London, 1658.
8. R. Gough, *Camden's Britannia*, J. Stockdale, London, 1806.
9. R. E. D. Clark, Will-o'-the-wisp, *Sch. Sci. Rev.*, 1942, **23**, 138–147.
10. M. Lister, An Account of the Burning of several Hay-ricks by a Fiery Exhalation or Damp: And of the Infectious Quality of the Grass of Several Grounds. From the same Ingenious Person, *Phil. Trans. R. Soc.*, 1694, **18**, 49–50.

11. W. Pryce, *Mineralogia Cornubiensis*, T. Phillips, 1778.
12. E. H. Rudkin, Will o' the wisp, *Folk-lore*, 1938, **46**(8), 49–50.
13. A. A. Mills, Will o' the wisp, *Chem. Br.*, 1980, **16**, 69–72.
14. W. Derham, Of the Meteor called the *ignis fatuus, from observations made in England, by the Reverend Mr. W. Derham F.R.S. and others in Italy*, *Proc. R. Soc. London*, 1730, **36**, 204.
15. A. Baker-Blocker, T. M. Donahue and K. H. Mancy, Methane flux from wetland areas, *Tellus*, 1977, **29**, 245–255.
16. H. Milne-Edwards, *Leçons sur la physiologie et l'anatomie comparée de l'homme et des animaux, Production de lumière par les animaux*, Paris, 1863, vol. 8, pp. 93–120.
17. G. Tsubota, Phosphate reduction in the paddy field I, *Soil Sci. Plant Nutr.*, 1959, **5**, 10–15.
18. I. Devai and P. D. Delaune, Evidence for phosphine production and emission from Louisiana and Florida marsh soils, *Org. Geochem.*, 1995, **23**, 277–279.
19. J. R. Burford and J. M. Bremner, Is phosphate reduced to phosphine in soils?, *Soil Biol. Biochem.*, 1972, **4**, 489–495.
20. G. Gassmann and D. Glindemann, Phosphane (PH_3) in the biosphere, *Angew. Chem.*, 1993, **32**, 761–763.
21. R. O. Jenkins, T. A. Morris, P. J. Craig, A. W. Ritchie and N. Ostah, Phosphine generation by mixed- and monoseptic cultures of anaerobic bacteria, *Sci. Total Environ.*, 2000, **250**, 73–81.
22. A. A. Mills, Will o' the wisp revisited, *Weather*, 2000, **55**, 239–241.
23. M. Minnaert, *The nature of Light and Color in the Open Air*, Dover, New York, 1954.
24. S. Kondo, K. Tokuhashi, H. Nagai, M. Iwasaka and M. Kaise, Spontaneous ignition limits of silane and phosphane, *Combust. Flame*, 1995, **101**, 170–174.
25. J. D. Barry, *Ball Lightning and Bead Lightning*, Plenum Press, New York & London, 1980.
26. P. Hubert, Tentative pour observer la foudre en boule dans le vosinage d'eclaires declenches artificiellement, Rapport DPH/EP/*76/349, 5 Mai 1075*, 1975.
27. D. Cooper and R. Cooper, On the luminosity of the human subject after death, *Philos. Mag.*, 1838, **12**, 420–426.
28. W. A. Dutt, *Wild Life in East Anglia*, Methuen & Company, London, 1906.

29. H. W. Robinson, Luminous owls and the laying of ghosts, *Field*, 1930, **155**, 230.
30. E. N. Harvey, *A History of Luminescence: From the earliest times until 1900*, American Philosophical Society, Pa., 1957.
31. C. A. Sinker, The North Shropshire Meres and Mosses, Field Studies, Publication E2, Headley, London, 1962.
32. E. Greenwood, *The Flora of North Lancashire*, Palatine Books, Lancashire, 2012.

Subject Index

Page numbers in *italics* or with a suffix T indicate a figure or table respectively. Roman numeral page numbers are at the front of the book; those in **bold** refer to terms in the glossary.